Random Dot Stereograms

By Andrew A. Kinsman

The front and back covers were rendered in PostScript from programs written by the author. The Spiral of Archimedes on the cover was generated from the gray-scale figure presented on the title page. The chair seat weave on the back is a relatively short program. When previewed on a computer screen it is entertaining to watch as the computer weaves each tile in succession.

October 1992   First printing
June 1993   Second printing

10 9 8 7 6 5 4 3

ISBN 0-9630142-1-8

Published by Kinsman Physics, P.O. Box 22682, Rochester, N.Y. ,14692-2682.
Additional copies of this book are available from the publisher for $17.00 post paid

A 3.5 inch floppy disk with the source code for programs listed in this book is available from the publisher for the fee of $10.00 U.S.

This book was typeset with Microsoft Word on a 486 PC clone. Some antiquated paste-up techniques were used due to disk space limitations. Fonts used include Times Roman for most text, Courier for computer programs, and Helvetica for section headers.

TRADEMARKS:

Ghostscript	is a registered trademark of Aladdin Enterprises.
GW-BASIC	is a registered trademark of the Microsoft Corporation.
LASAR	is a registered trademark of Perceptron, Inc.
LaserWriter	is a registered trademark of Apple Computer, Inc.
Linotronic	is a registered trademark of the Linotype Company.
MS-DOS	is a registered trademark of the Microsoft Corporation.
OpenWindows	is a registered trademark of Sun Microsystems.
PostScript	is a registered trademark of Adobe Systems.
QMS	is a registered trademark of QMS, Inc.
UNIX	is a registered trademark of AT&T Bell Laboratories.
X Window System	is a registered trademark of the Massachusetts Institute of Technology.

To my father,

for raising me to be an engineer,

and to my wife,

for accepting me in that condition.

Contents

Preface

This book represented a significant personal challenge. With it came many difficult decisions about ink, paper, font usage, and page formats. For some it will be a book of artwork, filled with mesmerizing pages of almost holographic intrigue. Others will purchase it for the programs and examples that exhibit the capabilities of random dot stereograms. In an effort to keep from mixing the technical with the esthetic, the text is often separated from the finished stereograms. This may cause a lack of continuity for the technical reader, but will make it easier to scan the stereograms in pursuit of a favorite. I have attempted to keep the tops of stereograms on the right edge of the right-hand pages. Stereograms that can be viewed inverted are more likely to be presented on a left-hand page. This will enable the viewer to turn the book 90 degrees counterclockwise and riffle through the stereograms, viewing those on the left-hand pages with the sight marks at the bottom (inverted), and in so doing, minimize the rotation problem. Presenting the stereograms in a rotated fashion has two benefits. First the pages remain flatter in the horizontal direction and second, sighting marks can be placed closer to the edge, allowing a distant object to be viewed more easily while one attempts to fuse the stereogram. I hope that the final results are as entertaining as the originals received by the printer.

It is believed that viewing these images cannot damage your eyes, but if you should happen to experience eye strain from the process, please set the book aside for a few days. Fortunately, several texts that include quantities of stereograms exist, and therefore set a precedence for any possible liability of my presenting more. In any event, the author is in no way forcing the reader to enjoy this book.

Chapter 1   Introduction

A random dot stereogram (RDS) is a collection of black and white picture elements (pixels) scattered in a seemingly random fashion over the surface of a page. Previously the patterns were often printed in two disjoint rectangles, but more recently the two images have been interwoven into a single rectangle. Either way the results are the same, an intriguing stereoscopic effect when viewed correctly. Several texts have been written on the use of RDSs for the study of stereopsis. This book will not be added to that collection. The author is not qualified in any way to discuss the brain's inner mechanisms for the assimilation of stereoscopic images. This book may make you far more aware of what is going on behind the eye. You should take some time to speculate on how each eye is being fed a meaningless image, and how this image can be transformed into a three-dimensional image that seemingly rises from a flat surface. It is quite possible that the process is responsible for the greater portion of our depth perception. It is my hope that this book will inspire others working in the perception field to make their own stereograms, and possibly in so doing, increase the total knowledge of stereopsis.

A few major advantages of stereograms over other three-dimensional viewing techniques are that they can be easily presented on a computer screen, require no special tools for the trained viewer, and can survive the all-too-common electrostatic copiers and FAX machines. They can also be printed on commonly available laser printers.

Disadvantages include the lack of color, or even continuous gray-scale information. Any attempt to add color only seems to distract the viewer. In fact, the best random dot stereograms appear completely random to the casual observer, with no indication of the underlying structure.

The history of the stereogram is a bit elusive. It appears to be intertwined with anaglyphs, lenticular photographs, and stereoscopic photographic techniques. Charles Wheatstone described stereoscopy in 1832. In 1851 the London Society of Arts held the Crystal Palace Exhibition, which six million people attended and potentially witnessed Sir David Brewster demonstrate the stereoscope. Stereoscopes became popular as a result. Kahn (1967), in *The Codebreakers*, references an article by Herbert C. McKay, written in the late 1940s, on how to manufacture simple stereograms with a typewriter for encryption purposes. This process will be illustrated later in the chapter on ASCII stereograms. My father observed lenticular photographic images for sale in Paris shortly before World War II. Julesz (1971) describes photographic techniques producing random dot stereograms in use in the early 1950s. History seems to have recorded no particular inventor of stereograms. It is quite probable that soon after parlor-style stereoscopes became popular someone took a photograph of a camouflaged hunter with a stereo camera. The subject in the resulting picture might be difficult to identify. Viewed stereoscopically with the rest of their collection, the subject would become obvious. Since Julesz, in 1960, was the first to employ a computer to generate random dot stereograms, many would consider him the person most responsible for their popularity today.

Chapter 2   Viewing Stereograms

There is a strong association between the muscles that control the convergence angle of your eyes and those that adjust the shape of the lenses in your eyes and focus an image on your retina. These two sets of muscles usually work together, trained by years of experimentation. The coordination of these two muscle sets must be decoupled in order to view stereograms without optical aids.

To view a stereogram each eye's line of sight must initially be through one of the (hopefully available) sight alignment arrows. This can be performed in one of many ways. The first and preferred technique is called "far-eyed", or "wall-eyed", viewing. In far-eyed viewing you initially train your eyes on a distant object. While doing so, slide the stereogram in front of you at a distance where the line of sight of each eye happens to be through each of the alignment arrows. Some find it easier to look over the top of the stereogram at a distant object, and eventually glance down into the stereogram when the alignment arrows finally fuse.

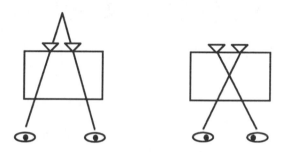

Figure 2.1 - Far-eyed and x-eyed stereogram viewing.

When done properly the two arrows appear as three, with the center arrow being much darker than the outer two. All three may initially appear fuzzy, but if you relax your eyes, they may come into focus along with other objects in the picture. The objects that do appear are best described as if they had been surface coated with a painted texture. A flat surface will appear behind the paper with an object rising out of it. Figure 2.2 contains the output of the program RDS.BAS, and is a practice single-image, random dot stereogram. It contains a disk above a flat background.

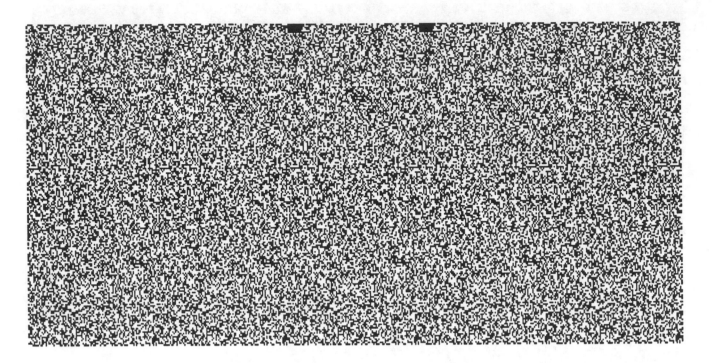

Figure 2.2 - Disk over flat background.

If you have tried far-eyed viewing and failed, now attempt what is called "cross-eyed" or "x-eyed" viewing. X-eyed viewing also enables each eye's sight line to proceed through a sight arrow, but in this instance the lines cross in front of the paper instead of behind. Most people find it helpful to hold their finger halfway between the paper and their nose. While looking at your finger, focus on the paper behind. As before, the two arrows become three fuzzy ones, which eventually may focus along with the rest of the image. This will make most of the stereograms in this book appear as hollow molds. The majority are printed for far-eyed viewing. Instead of objects that jump out at you as intended, they will be depressed into a surface. The flat surface in this case will be above the paper.

Perhaps you are skeptical that this book is a hoax. If so, you are probably not alone. I will assume your attempts have so far failed. Fear not, there are still many tricks available. One is a variant on far-eyed viewing to trick you into latching onto the depth information stored in the stereogram and not automatically converging on the paper. Place the stereogram close to your nose, and slowly draw it farther away to arm's length, straight out from your nose. By starting with the stereogram at a distance too close to focus on initially, your natural tendency to seek at infinity for a "lost horizon" might assist you. Many individuals obtain instant viewing success with this technique.

By now you will need some proof that there are actually objects hidden in these pictures. To find them you can photocopy the stereogram, first onto a transparency, then onto another white sheet of paper. Doing so makes sure that both have been transformed identically in scale by the same copier and are suitable for overlaying one upon the other. By slightly shifting the transparency left and right from the position where the alignment dots overlap you may be able to see a hint of the subject of the stereogram. This is especially noticeable on ASCII two-level stereograms. Some find it helpful to cut the fusion alignment marks from the top of the transparency, overlaying them directly on the center of the paper copy. Figure 2.3 shows how the transparency overlay technique exposes for the toroid stereogram. Here the stereogram is printed once, then a second time shifted right a distance equal to the spacing between the sight marks. No three-dimensional effect is present in this illustration, but only an idea of how the shape should look, plus proof that more than random noise is present.

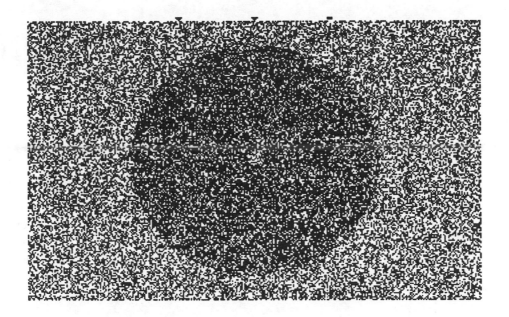

Figure 2.3 - Transparency overlay print of toroid (not fusible).

It is possible to view random dot stereograms using two dove prisms. Placed back to back, each will mirror the image presented. As long as each is mirrored, no damage is done. While trying not to rotate the prisms out of alignment, steer your vision path around to cause your left eye to be looking straight at the left sight, and the right eye to be looking directly at the right sight. Again an object may appear in the image. Dove prisms can be obtained from Edmund Scientific, and the piles of discarded optics that have an affinity for hamfests. Hamfests are special flea markets frequented by amateur radio enthusiasts for trading the necessities of their hobby. The general public is always welcome. Usually you can visit the house in your neighborhood that has too many antennas for details on when and where to find a hamfest. When dove prisms are available they rarely cost more than a few dollars each.

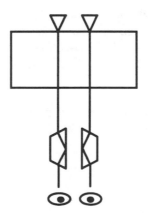

Figure 2.4 - Dove prism viewing technique.

An alternative to two dove prisms is a wedge prism used to divert the sight line of only one eye, or two thinner wedges, one used in front of each eye. Again, hamfests are the best place to find these.

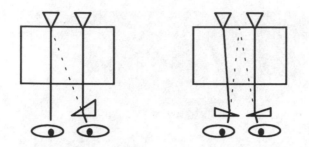

Figure 2.5 - Single/double wedge prism viewing techniques.

Another alternative to a wedge prism is a simple piece of mirrored glass. If a double-image random dot stereogram (DIRDS) is printed with the second box mirrored over, the two images can be fused by placing a mirror vertically between them. Best is a front-silvered mirror, but back-silvered mirrors work with only the slight addition of some ripple to the reflected image. This technique won't help for the rest of the book, but will provide a way of knowing what you are looking for and is every bit as effective as the following option.

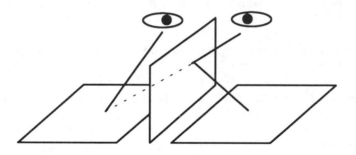

Figure 2.6 - Technique for viewing a mirrored DIRDS.

A final resort is to find anaglyphs of random dot stereograms. These are presented in the back of Julesz's book, and a pair of the (half-red/half-green) glasses required to view them is tucked inside the back cover. A random dot stereogram anaglyph postcard can be ordered from Mr. Balogh (see Chapter 10). This book is printed without color, so it was impossible for me to include one here. Sadly, some individuals may find that even these produce absolutely no stereoscopic effect.

It does appear that with minor training approximately 90% of the population is able to view random dot stereograms, either with or without optical aids. The remaining 10% are either unwilling, or unable, to see the effect.

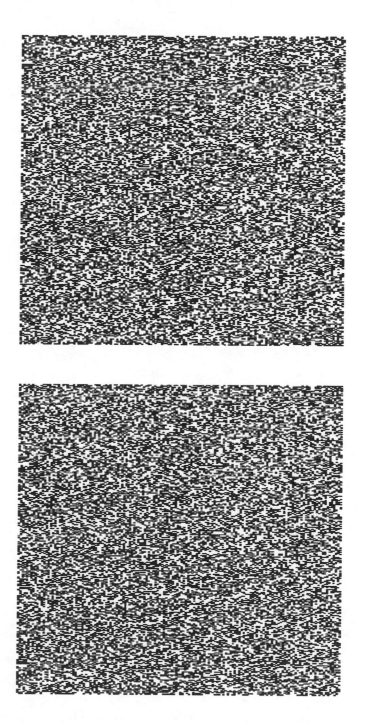

Figure 2.7 - DIRDS suitable for viewing with a mirror (mirror is required).

Chapter 3 How Do They Work?

Years of experience correlating the visual information that our eyes pass to our brain has trained us to transform this information into a perception of depth. Sometimes we know the relative size of an object, and its distance is determined from the fact that it seems much smaller than when it is at arm's length. At other times we determine the distance from the convergence angle of our eyes as they steer toward an object. Alternatively, we may obtain distance information for the lens muscles that focus the image on the retina of our eyes. Remaining is the most important clue that forms the basis for this book. Possibly you have awakened staring at a wall covered with a narrow repeating pattern of wallpaper. You are certain that you have the wall in focus, but are unable to determine its distance. This effect, aptly called the "wallpaper" effect, is caused by each eye's sight path piercing one of the patterns. As you attempt to correlate the parallax of objects in the scene, the brain is content that the convergence angle of the eyes is correct, and focus is correct, so the wall must be at the distance computed. You reach out and touch the wall to find it only half the distance you had expected. This correlation of the objects in the field of view helps us determine depth perception, and is what stereograms are all about. Each eye receives meaningless noise as an image, while the brain manufactures a three-dimensional object that unveils itself for the viewer.

Look somewhat cross-eyed at the pattern in Figure 3.1. Use any pair of dots as sight marks. When you have fused it as described in the previous chapter you will find that some of the eyes pop out, and others are depressed. The shifting in the printing of the circles and spots causes the effect. Depending on the amount of shift, many different apparent elevations are possible. Figure 3.2 shows the depths that various shifts could represent.

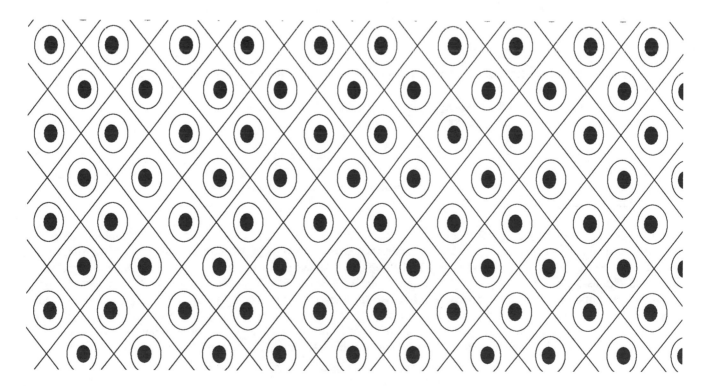

Figure 3.1 - Diamond wallpaper with eyes.

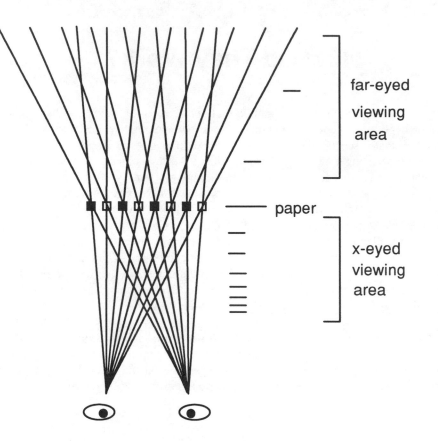

Figure 3.2 - Crossing points of viewing far and x-eyed stereograms.

A random line stereogram or a picture stereogram may yield insight into the importance of correlating the relative shifts of objects in each eye's field of view. Note that objects to the right of the left eye's sight line, and left of the right eye's sight line, are understood to be closer to you than objects in the direct line of each eye. Try winking one eye then the other, and examine how the picture changes. Figure 3.3, prepared as a random line stereogram, may illustrate more clearly how the shift is produced in a random dot stereogram.

All stereo viewing techniques exploit the brain's ability to detect subtle shifts in the images each eye receives. This parallax is interpreted as a three-dimensional image. Each eye is fed a slightly different image, and the organic computer goes to work discovering where the information differs in its endless task of mapping the distances to the objects around us.

Now that you know how to fuse random dot stereograms, you will be able to fuse other stereo photographs. The next time you use your camera, take two pictures. Take one, then step a foot to the right and take a second. If the subject is stationary, for example a landscape scene, you will be able to fuse the two resulting photographs. A close look may reveal that clouds have actually moved during the period between shots. This may cause some visual confusion. Peruse your photograph collection and see if you have two pictures taken in sequence from the window of a train, or an airplane. Place them side by side and view them x-eyed. Switch them around and fuse them again, to find the correct order. Whenever you see duplicate photos or coupons, try to fuse them and see if some anomaly in exposure or printing accidentally produced a stereo pair. Pizza coupons in the back of telephone books are a common example.

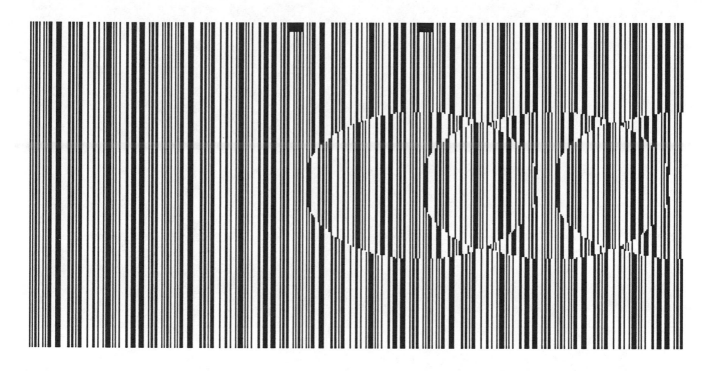

Figure 3.3 - Random line stereogram.

Chapter 4   ASCII Stereograms

One simple type of stereogram is the ASCII stereogram. Examples are included in this book primarily because some people find them easier to view. Most have only two elevations, a background and a foreground. They can be manufactured with a simple text editor, and the program listed in Appendix A.

First a file of two types of characters is prepared with a simple ASCII editor. In the DOS environment, the utility "edlin" will work, but an editor that enables you to view an entire screen is far better. Start with a screen filled completely with the background character, and then modify only those characters that will become the foreground. Once this pattern is prepared, send it through the program mentioned, resulting in an ASCII stereogram. This command was used to produce Figure 4.2:

```
%stereogram <ascii.msk >ascii.lst
```

```
------------------------------------------------------------------
------------------------------------------------------------------
------------------------------------------------------------------
-----------------------------------#------------------------------
-----------------------------------#------------------------------
----------------------------------###-----------------------------
----------------------------------###-----------------------------
---------------------------------#####----------------------------
---------------------------------#####----------------------------
--------------------------------######----------------------------
----------------------#####################-----------------------
-----------------------##################-------------------------
------------------------##############-----------------------------
-------------------------#############----------------------------
--------------------------##########-------------------------------
---------------------------##########-----------------------------
---------------------------##########-----------------------------
--------------------------############----------------------------
--------------------------#####---#####---------------------------
-------------------------#####-----#####--------------------------
------------------------###---------###---------------------------
-----------------------###-----------###--------------------------
----------------------#-------------#-----------------------------
------------------------------------------------------------------
------------------------------------------------------------------
------------------------------------------------------------------
------------------------------------------------------------------
------------------------------------------------------------------
------------------------------------------------------------------
------------------------------------------------------------------
```

Figure 4.1 - ASCII stereogram mask.

```
                              X                      X
FOZEPDKDNQLHFEJDFOZEPDKDNQLHFEJDFOZEPDKDNQLHFEJDFOZEPDKDNQLHFEJDFOZEPDKDNQLHFEJD
ZKSCCFYKDXNLORWVZKSCCFYKDXNLORWVZKSCCFYKDXNLORWVZKSCCFYKDXNLORWVZKSCCFYKDXNLORWV
FWAVBMYQCLXJRGNTFWAVBMYQCLXJRGNTFWAVBMYQCLXJRGNTFWAVBMYQCLXJRGNTFWAVBMYQCLXJRGNT
QHVUOWGRTUFHNBDTQHVUOWGRTUFHNBDTQHVUOWGRTUFHNBDTHZVUOWGRTUFHNBDTHZVUOWGRTUFHNBDT
FQATQRYEQHXXUQPDFQATQRYEQHXXUQPDFQATQRYEQHXXUQPDQMATQRYEQHXXUQPDQMATQRYEQHXXUQPD
MSKSJDOONCSSVRZNMSKSJDOONCSSVRZNMSKSJDOONCSSVRZMSKSSJDOONCSSVRZMSKSSJDOONCSSVRZM
SFLSJBAHAWXSALESSFLSJBAHAWXSALESSFLSJBAHAWXSALESFLVSJBAHAWXSALESFLVSJBAHAWXSALES
WDBLSQPKIMDJZXDEWDBLSQPKIMDJZXDEWDBLSQPKIMDJZXEWDBLISQPKIMDJZXEWDBLISQPKIMDJZXEW
WQMWXOUWGUKKMFJQWQMWXOUWGUKKMFJQWQMWXOUWGUKKMFQWQMWIXOUWGUKKMFQWQMWIXOUWGUKKMFQW
WKYNWIZZTYXXEHTUWKYNWIZZTYXXEHTUWKYNWIZZTYXXETUWKYNWVIZZTYXXETUWKYNWVIZZTYXXETUW
RTKLQSGADUHOMSYRTKLQSGADUHOMSYRTKLQSADUHOMSMYRTKLQSADUHOMSSMYRTKLQSADUHOMSSMYR
ZWBYWWYVTEEOZAFUZWBYWWYVTEEOZAFUZWBYWWYTEEOZAFUZWBYWWYTEEOMZAFUZWBYWWYTEEOMZAFUZ
TMEBOJVWXKQLIIMHTMEBOJVWXKQLIIMHTMEBOJVWXQLIIMHTMEBOJVWXLQLIIMHTMEBOJVWXLQLIIMHT
MFIKABPGSQIZKUSZMFIKABPGSQIZKUSZMFIKABPGSQZKUSZMFIKABPGZSQZKUSZMFIKABPGZSQZKUSZM
TNIRLIBBTLKGSVUZTNIRLIBBTLKGSVUZTNIRLIBBTLKSVUZTNIRLIBDBTLKSVUZTNIRLIBDBTLKSVUZT
FBHWIBOQAALTZKMNFBHWIBOQAALTZKMNFBHWIBOQAALZKMNFBHWIBOSQAALZKMNFBHWIBOSQAALZKMNF
DYCGAWUKEOHYONFPDYCGAWUKEOHYONFPDYCGAWUKEOHONFPDYCGAWUWKEOHONFPDYCGAWUWKEOHONFPD
DXXRQXUBQTNFGHKHDXXRQXUBQTNFGHKHDXXRQXUBQTFGHKHDXXRQXUBKQTFGHKHDXXRQXUBKQTFGHKHD
RHINTOBCORPWHLZGRHINTOBCORPWHLZGRHINTOBCORWHLZGIRHNTOBCTORWHLZGIRHNTOBCTORWHLZGI
YINBYIOFRYQYKJHSYINBYIOFRYQYKJHSYINBYIOFRQYKJHWSYINYIOFRCQYKJHWSYINYIOFRCQYKJHWS
IZGWRWSAJUIUPHCEIZGWRWSAJUIUPHCEIZGWRWSAJIUPIHCEIZGWRSAJCIUPIHCEIZGWRSAJCIUPIHCE
MZIFXYFJHODFQLEXMZIFXYFJHODFQLEXMZIFXYFJODFHQLEXMZIFXYJODXFHQLEXMZIFXYJODXFHQLEX
VCXGYGANPERXHFYRVCXGYGANPERXHFYRVCXGYGANENRXHFYRVCXGYGANNXRXHFYRVCXGYGANNXRXHFYR
PSGYHJDZUHYEFVIEPSGYHJDZUHYEFVIEPSGYHJDZUHYEFVIEPSGYHJDZUHYEFVIEPSGYHJDZUHYEFVIE
CGKCVBHOZQVZHIXYCGKCVBHOZQVZHIXYCGKCVBHOZQVZHIXYCGKCVBHOZQVZHIXYCGKCVBHOZQVZHIXY
DDWKPZLLIKRPFZUHDDWKPZLLIKRPFZUHDDWKPZLLIKRPFZUHDDWKPZLLIKRPFZUHDDWKPZLLIKRPFZUH
HGMEIVUJLQIUAFSDHGMEIVUJLQIUAFSDHGMEIVUJLQIUAFSDHGMEIVUJLQIUAFSDHGMEIVUJLQIUAFSD
LOPAPBNXLFNSEHAOLOPAPBNXLFNSEHAOLOPAPBNXLFNSEHAOLOPAPBNXLFNSEHAOLOPAPBNXLFNSEHAO
PMSXJPGUFPOFWGLHPMSXJPGUFPOFWGLHPMSXJPGUFPOFWGLHPMSXJPGUFPOFWGLHPMSXJPGUFPOFWGLH
WAJLBXKMCXFIGHWWWAJLBXKMCXFIGHWWWAJLBXKMCXFIGHWWWAJLBXKMCXFIGHWWWAJLBXKMCXFIGHWW
```

Figure 4.2 - Resulting ASCII stereogram.

In the UNIX environment the commonly available tool "bitmap" will allow you to rapidly make X Windows bitmaps. A subsequent program "bmtoa" can then be invoked to convert this xbitmap into an ASCII file containing these same "#" and "-" characters, as the two elevations in an "elevation map". This elevation map can then be passed through the ASCII stereogram program to make the stereogram. Elevation maps used later on will be composed not of ASCII characters, but instead the binary values 0 to 255. This will permit more elevations to be portrayed by each byte, but with the side effect of making the elevation maps more difficult to manipulate with a computer.

One particularly interesting version of the ASCII stereogram program produces ASCII stereograms, and is itself an ASCII stereogram, due to careful replication of the program in the comments on the right side. Viewing it (x-eyed as the separation is too great) causes its title to jump toward the viewer, and the center to be dimpled inward. This is an example of how a stereogram could be used to hide a coded message as described by McKay. A pair of pages filled with either random words or letters could have some adjusted so they would stand out.

```
              /*/                              /*/
#include <math.h>/*STEREOGRAMSX/**#include <math.h>/*XSTEREOGRAMS/**/
#include <stdio.h>/*IN ASCII!$%/**#include <stdio.h>/*$IN ASCII!@/**/
main(){char m[80],a[80],e,j;for/**main(){char m[80],a[80],e,j;for/**/
(;gets(m)!=NULL;puts(a))for(a[ /**(;ges(m)!=NULL;puts(a))fork(a[ /**/
80]=e=j=0;j<80;j++)a[j]=e&&m[j-/**80]==j=0;<80;j++)a[j#]=e&&rm[j-/**/
16]!='#'||j<16?e=0,random()%26+/**16]!'#'||j<16?e=0,random()z%26+/**/
'A':a[j-16+(e=m[j-16]=='#')];}  /**'A':a[j-16+(e=m[j-16]=='#')];}  /**/
```

Figure 4.3 - Stereogram of STEREOGRAM.C. © Greg Alt 1992.

Chapter 5   Producing Random Dot Stereograms

It is presumed at this point that you are in possession of an elevation map. The following chapter will describe different ways of manufacturing them. An elevation map is then easily transformed into a random dot stereogram by the algorithm illustrated in Figure 5.1.

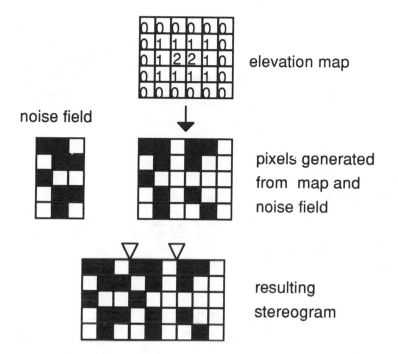

Figure 5.1 - Transforming an elevation map into a stereogram.

The algorithm is quite simple. An area of pixels must be generated from a noise field and an elevation map. The method of generation involves choosing an offset. The offset chosen is the spacing between the sight marks in the stereogram. The offset is a particular number of pixels on the output device. To determine the color of a pixel in the generated field you merely count "offset" pixels to the left and "elevation" pixels back to the right. The color of the pixel you come upon defines the color of the pixel you are trying to define in the generated field. Repeat from left to right for all columns and rows in the generated field. For this algorithm to work, at least "offset" pixels of random colors must be previously assigned to the noise field at the left, as you will count left into this field for initial pixels on each line of the stereogram. A wider noise field will be required if you decide to use negative elevations.

The program RDS.BAS, shown in Appendix B, is a simple example of how to produce a single-image random dot stereogram on a PC computer using GW-BASIC. It follows the algorithm previously described. A slight modification, noted in its comments, will convert it into a program that generates random line stereograms. The output of RDS.BAS has already been exhibited as Figure 2.2.

A sequel program, DIRDS.BAS, also presented in Appendix B, produces traditional dual-image random dot stereograms. A slight adjustment to it will produce DIRDS suitable for viewing with a mirror.

A more flexible method for the production of stereograms is to build small programs that generate elevation maps, and employ a second program to convert the elevation maps into stereograms.

Appendix B contains examples of programs that perform this sequence. The program DIAMOND.BAS generates the elevation map called DIAMOND.ELV. It can be formatted into a stereogram with the program RDS-ELV.BAS. Take care in the MS-DOS environment not to generate elevations of the value 26, as this character value is used as the end-of-file character. The elevation maps are stored as character files to keep their sizes to a minimum, one character per elevation point in the grid.

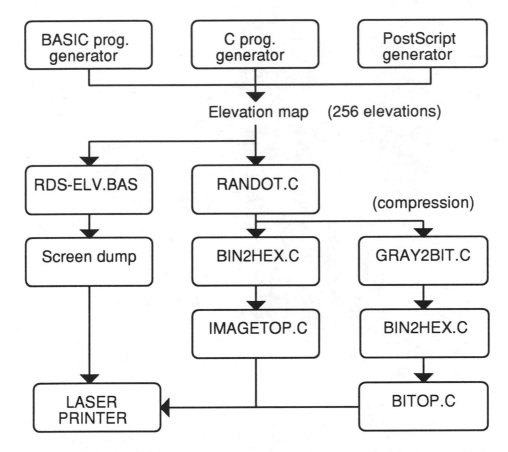

Figure 5.2 - Flow diagram used to prepare stereograms in this book.

Without significant computing resources, the BASIC program sequence illustrated by DIAMOND.BAS and RDS-ELV.BAS is probably the easiest way to produce stereograms. The "print screen" button will dump them to a dot matrix printer. A few lines can be tacked on the end of RDS-ELV.BAS to save the screen into a file, enabling you to ship the resulting image to a laser printer if one is available.

Use the "merge" command in BASIC to append these lines and produce a screen dump of BASIC-generated screens. The utility BIN2HEX.C, followed by editing IMAGETOP.PS to the beginning, will allow you to convert the file produced (SCRNSAVE) to a PostScript file.

```
1000 OPEN "scrnsave" FOR OUTPUT AS #2
1010 FOR Y=0 TO 199
1020 FOR X=0 TO 639
1030 VA=255
1040 IF POINT(X,Y) = 1 THEN VA=0
1050 PRINT#2,CHR$(VA);
1060 NEXT X
1070 NEXT Y
```

After preparing the necessary elevation map, some of my first stereograms were generated using RANDOT.C with the following command:

```
%randot 800 800 0 <input_elevation_map | bin2hex >stereogram.ps
```

The resulting stereogram was 1000 pixels wide and 800 pixels high. It then was edited to include a header. See IMAGETOP.PS in Appendix D.

In an effort to save space, because the household PC-AT clone has only a 20-megabyte disk, many of the stereograms were converted, from the inefficient form that RANDOT.C leaves them in, to bitmaps. Bitmaps are slightly more difficult to understand in PostScript. The program GRAY2BIT.C performed this important transformation, with BITOP.PS used as the header, instead of the less complicated IMAGETOP.PS. Here is the command sequence used:

```
%randot 800 800 0 <input_elevation_map | gray2bit 1000 | bin2hex >stereogram.ps
```

The resulting stereogram was again 1000x800 pixels, but occupied about 1/8 the disk space of the original. Each original byte output from RANDOT.C was 0 (black) or 255 (white), now reduced to a single bit representing black or white. BITOP.PS also prints sight marks and a copyright notice on the finished stereogram.

For those interested in the use of PostScript to generate RDSs, please see DOTOP.PS in Appendix D. It optionally takes what would otherwise be a gray-scale image and interprets it as either an elevation map or an image. This program is slow, but provides the equivalent of a random dot operator in the PostScript language. It is also a convenient way to produce black and white prints showing the subject of the stereogram.

You may wish to seek out a professional print shop with your best artwork. Linotronic raster image processors (RIPs) will readily absorb PostScript, but it may cost you about $10 per print. The higher resolution models will generate 2400 DPI for more spectacular RDSs. Be wary, prices increase with resolution. Some shops will not wait around for prints that take two or three days to render. The earlier Linotronic RIP-1 and RIP-2 devices aren't known for their speed at executing numerically intensive PostScript. Some of their newer RIPs are significantly faster. Seek out the smaller shops, which may permit your job to run overnight.

If you wish to generate complex tiling patterns with a numerically intensive PostScript program, it may be better to use one of the versions of PostScript-like languages that can render the page into your computer's memory. Ghostscript is one of these programs. It is distributed by Free Software Foundation, Inc. and runs in both the DOS and UNIX environments. Several similar versions exist. This pixel map is then stored and prepared as a PostScript image. The resulting image can be rapidly gobbled up by most commercial PostScript devices, saving many hours of rendering time at the print shop. One good reason for doing just the opposite is that some images are far too large to transport to a print shop. Most shops commonly accept 80-megabyte removable hard disks. Some optics designs exceed that by an order of magnitude, but can be expressed mathematically in relatively few lines of PostScript. See Appendix D for an example of a PostScript program that computes, internal to the printer, a Mandelbrot fractal plot. It makes an excellent RIP performance test program.

It is possible to combine the DOTOP.PS program, which renders an elevation map directly into a stereogram, with the Mandelbrot program. The results are about four pages of PostScript code that both mathematically compute a topology and convert it into the resulting stereogram. This can produce extremely dense RDSs manufactured completely internal to the RIP.

Some laser printers have a problem with the program DOTOP.PS because they have a relatively poor random number generator. They will produce images that resemble Figure 5.3.

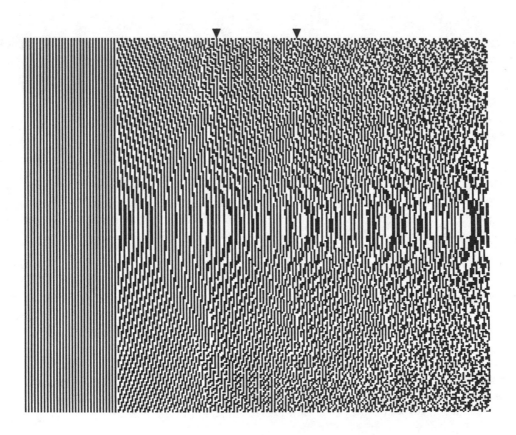

Figure 5.3 - DOTOP.PS output for laser printer with poor random number generator.

Chapter 6   Elevation Maps

One of the easiest ways to produce elevation maps for stereograms is from an equation. Dust off that old math book, and the chapter on cardiods will rekindle your interest. Any equation, plotted in an interesting way, will produce a data set that will yield results. Some mistakes can be more interesting than the intended results. With the popularity of fractals, it only seemed natural to experiment with fractal equations as generators for the elevation maps used in random dot stereograms.

Stereograms plotted from equations are shown on pages 41 through 65.

Other potential sources of elevation maps are databases. Topological databases are an excellent find. They can be even more fun if you convert them to the terrain format required for your favorite flight simulator. It is pure enjoyment to dogfight over the volcanoes of Mars.

Pages 66 through 72 contain stereograms produced from topological databases.

Other data files of topologies you might keep an eye open for are airfoil surfaces, CAD projections, geological files, and half hull data sets, used in yacht designing.

The language PostScript is one other excellent source of elevation maps. Programs are written to form images in this intriguing language. The images normally sent to PostScript printers can be easily designed with a little study. Some very simple programs can produce startling results. Gray-scale images in PostScript have a pixel of the value zero in memory for black, and the value 255 for white. This gives you 256 combinations, which for our needs means 256 different elevations. Even something as simple as one of the ZapfDingbat characters printed repetitively smaller and lighter can produce an interesting and usable elevation map. Figure 6.1 shows the PostScript program that generated the continuous stereogram on page 34.

Pages 73 through 90 contain stereograms generated from elevation maps manufactured by PostsScript gray-scale images.

If your computer has a scanner attached to it for scanning either pictures or text, it too can be the source of interesting elevation maps. While scanners can produce data quickly, it often is very irregular in texture. Appendix C has a program for smoothing data produced in this way. If the ROTATE90.C program is employed it can be smoothed in two directions. With several iterations of rotation and smoothing, good results can be obtained. As with the language PostScript, scanned images are often a single byte per pixel scanned, with 0 representing black and 255 representing white. Some scanned photographs translate directly into an elevation map of the subject, primarily because of creative lighting. Often this lighting is accidental. Photographs of clouds, hubcaps, gravestones, neon signs, and some microscope pictures fall into this category.

Pages 91 through 94 show stereograms generated from photographs.

There are some unusual devices for surface analysis and quality control used in industry. These devices are able to scan a surface and produce an X-Y-Z data file of the topology of the surface. Some will even produce a residual file of the actual surface color. The "Superhuman" was made with a device called a LASAR. Manufactured by Perceptron, Inc., this 3-D camera scans the scene with a laser, recording the beam echo time for each pixel.

Pages 95 to 96 show stereograms generated with X-Y-Z devices.

```
%!
%   AUGER.PS - continuous RDS example
%
%   By A. Kinsman 8/28/92
%
/turns           1    def       % number of turns
/radius {     128 } def         % initial radius coeff


clippath fill                             % black background
256 4 mul 128 sub 64 sub 128 translate    % go to center of field-fudge
90 rotate                                 % entry at top
4 setlinewidth
360 turns mul -.1 0 {
                dup                       % ang ang
                % x= r cos theta
                cos radius mul            % ang x-val
                % y= r sin theta
                exch
                sin radius mul            % x-val y-val
                0 0 moveto                % computer inner rad
                lineto stroke

                % whiter/brighter with each line
                currentgray 1 360 turns mul 10 mul div add setgray
        } for
showpage
```

Figure 6.1 - PostScript code for continuous RDS.

It is possible to build your own device to obtain the topology of a person or object. The subject is illuminated with horizontal stripes of light projected from a source that is raised 45 degrees from horizontal. A picture is taken from directly above, capturing the striped pattern. Since the geometry of the lighting is predetermined, analysis for topological information can be performed with simple equations. A frame grabber board makes this task even easier for those lucky enough to own one. Many computer-assisted assembly-line vision inspection processes employ a similar system. An inexpensive alternative is to tape a transparency of the photograph to your computer screen and use a mouse to digitize the image. You can also create topologies with your favorite paint utility. Fill in regions with lighter colors as the elevation increases. The resulting image can be stored and converted to a stereogram.

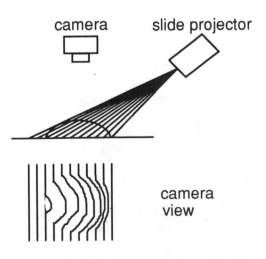

Figure 6.2 - Homemade topology gathering.

Chapter 7   For Gifted Viewers

The stereogram on the following page (Figure 7.2) was generated in rectangles. Each area is increasing in the number of pixels that are placed as a function of random noise instead of their correct elevations. It is designed to measure your ability to observe the effect. See how far down the page you can continue to find the numbers, which describe the percentage of pixels placed at their correct elevation in the stereogram.

It is possible to generate a random dot stereogram that has more than one surface in a particular area. In fact, as the resolution of the printing device increases, many layers are possible. The viewer's eye can be tricked into following a particular layer if the surfaces intersect at an angle. When two ramps are used, one descending toward the right side, and another descending toward the left, one of the ramps becomes almost unnoticeable if the upper portion of the stereogram contains elevation information for a single ramp. The process of generating multilevel, or ambiguous, stereograms is quite easy. The elevation map prepared must merely be alternating in one direction or another between two levels. Most patterns produce results. The following collection of BASIC programs will generate two- and three-level RDSs. See if you agree that the best pattern is one in which all the elevations for one surface exist on a single line with alternating horizontal lines sequencing between surfaces being portrayed. This minimizes pixel stretching in the finished stereogram.

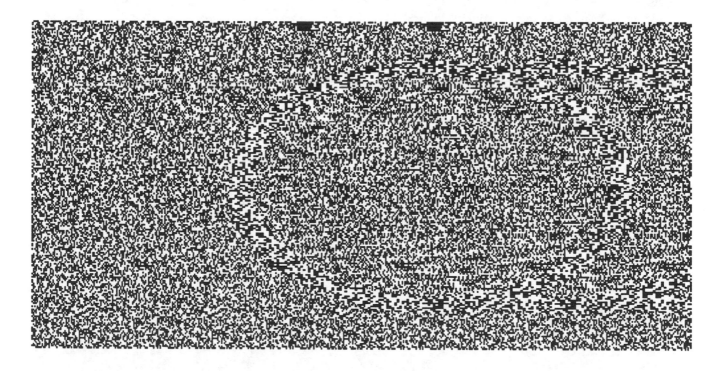

Figure 7.1 - TRDS1.BAS output (two surfaces alternate across rows.)

Figure 7.2 - 100%-95%-90%...65% RDS example rendered at 1200 DPI.

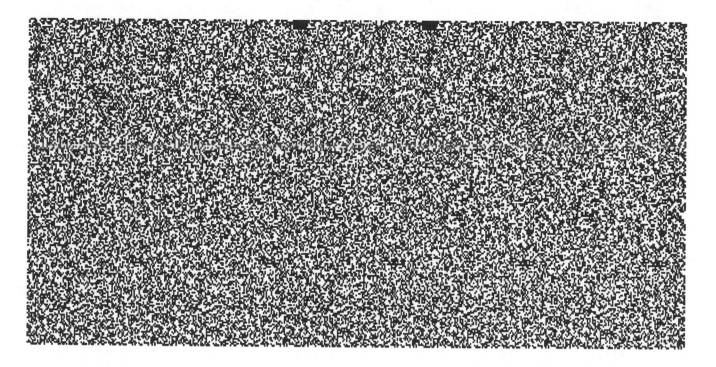

Figure 7.3 - TRDS2.BAS output (two surfaces alternate down columns.)

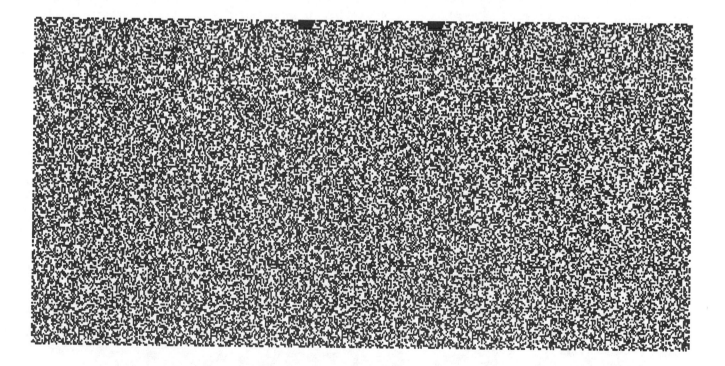

Figure 7.4 - TRDS3.BAS output (three surfaces alternate down columns.)

Figure 7.5 - TRDS4.BAS output (three surfaces alternate down columns.)

The initial field of a stereogram used to generate the rest of each horizontal line need not be random dots as presented up to this point. Dual-image stereograms with the random field filled in with triangles, lines, blotches, and even images produce perfectly acceptable results. These are often used by vision researchers to assist in their studies of depth perception. Ninio and Herlin (1988), and Slinker and Burton (1992), experimented with stereograms containing complex patterns in their initial noise fields. You can also produce these by modifying any of the programs listed to read in a previously prepared image into the beginning of each line. The algorithms that drop back "offset+elevation" pixels to obtain the current pixel value will still function.

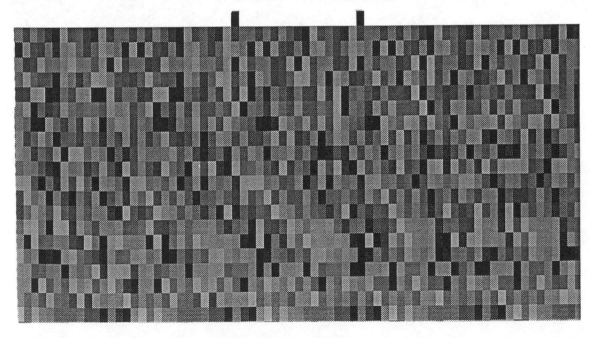

Figure 7.6 - Low-resolution example with "multicolor" pixels.

Figure 7.7 is manufactured with an image in the otherwise initial random field. Sight marks are often missing from this style, as the repeated image makes it relatively easy to fuse. Some even find this type the easiest to observe initially. It is produced by altering RANDOT.C in the area in which the initial field is produced. The program selects a byte from a prepared image file instead of randomly generating a white or black pixel.

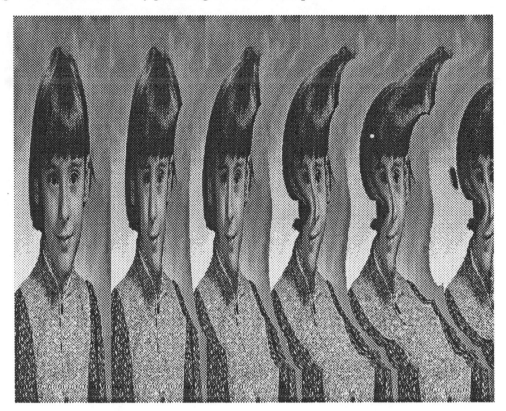

Figure 7.7 - Photograph warped by toroid.

Instead of including a photograph in the initial noise field, as in the last example, the image of a previously prepared RDS can be used. When rotated 90 degrees counterclockwise and inserted, with the same elevation map being used to generate a second stereogram, the illustration opposite Chapter 9 is produced. The viewer can use the sight marks to view one stereogram, or rotate the page 90 degrees clockwise and see the original that now represents the noise field.

Continuing one step further, if you produce a stereogram from a gray-scale image that is also interpreted as its elevation map, you have come full circle back to a stereo pair of images. Figure 7.8 illustrates this and should be viewed x-eyed to obtain the stereoscopic effect. The image on the left is repeated on the right with the pixels shifted a small percentage of their elevation.

One last oddity remains. Kontsevich (1986) describes a technique for making a series of tiles. These tiles, like steps in stairs, progress in a circular pattern toward the viewer. After each pass around the circle, the viewer's eyes become focused at a new elevation. This continues until eyestrain sets in, and the viewer reverses direction descending the steps.

Here is a similar continuous stereogram, Figure 7.9. Reminiscent of a boring tool, it is a ramp that climbs in a spiral toward the viewer. It is limited in height by the available space of random dots. Five fields of 256x256 are presented, the first filled with random bits, three generated with elevation zero, and the last with the object. The object must start and end in the same location, and climb to a height equal to the offset used to make the stereogram. After fusion is achieved, USING X-EYED TECHNIQUES ONLY PLEASE, cover up the sight marks and climb (clockwise) or descend (counterclockwise) the spiral found on the right-hand side of the stereogram. To help get you on track, the ramp starts at a fairly steep angle out of the top center of the circle observed.

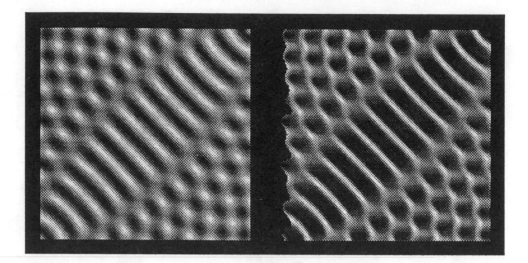

Figure 7.8 - Image warped with itself.

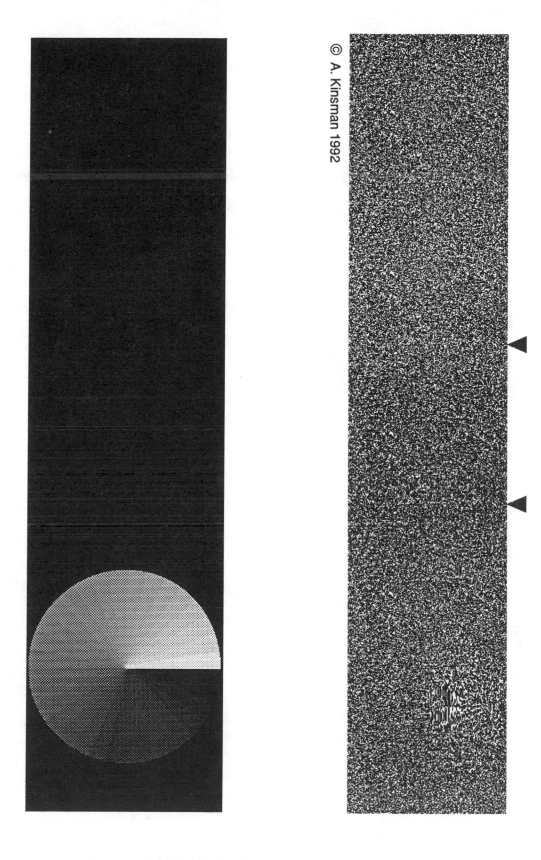

Figure 7.9 - Continuous spiral stereogram.

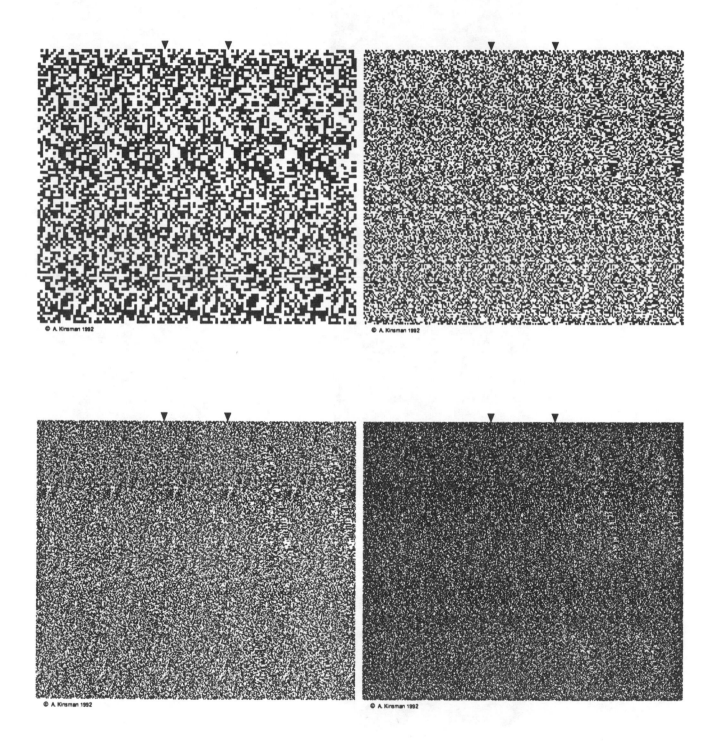

Figure 8.1 - LAMPSx4.PS a resolution study.

Chapter 8   Testing the Limitations

The effect of resolution was difficult for me to investigate. The difficulty being whether to study the resolution of the printing device or the density of the elevation map used to generate the stereogram. Figure 8.1 shows the effect of increasing the density of the elevation map. Shown are 100x100, 200x200, 400x400, and finally 800x800 point maps. Even here two variables are being modified, elevation map density and number of elevations presented. Without varying both, each would have half the relief of the previous image. Figure 8.5 shows the effect of continually reducing the scale of the image. In both cases the subjects appear continually smoother in surface texture, and darker, until the resolution of the printing device is exceeded. They then take on a twinkling, almost iridescent appearance, as pixel aliasing begins. Figure 8.5 is printed at 1200 dots per inch (DPI) on a Linotronic printer to ensure that the smallest can be viewed. In this case each image is produced from the same 800x800 point elevation map. Biasing the color selection of pixels in the noise field toward white will noticeably lighten higher-resolution stereograms.

It is easier to experiment with variations in offset. This can be done by photocopying one of the DIRDS and cutting it down the center to separate the two squares. With the squares separated this way, you can observe how the height of the object in the image is affected by different spacings between the squares. You will have to view them with the x-eyed technique to perform this experiment.

Here it is finally appropriate to discuss some of the mathematics associated with determining apparent elevations for a particular viewer. If an individual's eye spacing is measured, a proper viewing distance can be determined for the viewer. A knowledge of similar triangles and proportions is all that is required to understand Figure 8.3.

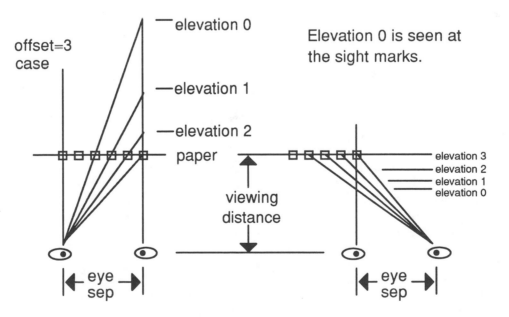

Figure 8.2 - Geometry of far-eyed and x-eyed viewing.

With viewing distance in inches, integer elevation numbers, and all other values in pixel units:

distance behind page surface = viewing distance * (offset - elevation#)/(eye separation - offset + elevation#)

distance in front of page surface = viewing distance * (offset - elevation#)/(eye separation + offset - elevation#)

Figure 8.3 - Equations for far-eyed and x-eyed viewing depths.

The BASIC program RDSDEPTH.BAS presented in Appendix E will display a table of apparent elevations for a particular viewer. This table, or an equation, could be used up front in the RDS generation sequence to determine the correct adjusted elevation for any original elevation. A sequel program RDSDPTH2.BAS generates the table for the x-eyed case. An example of the output of RDSDEPTH.BAS for a viewer with a 2.25 inch eye spacing, observing a 100 pixel per inch RDS, with an offset of 10 pixels, and a 24 inch viewing distance, would look like this:

```
base elevation is  1.116279  inches behind page
elev  1 is  0.1163 inches from elev 0, change is 0.1163
elev  2 is  0.2315 inches from elev 0, change is 0.1152
elev  3 is  0.3456 inches from elev 0, change is 0.1142
elev  4 is  0.4587 inches from elev 0, change is 0.1131
elev  5 is  0.5708 inches from elev 0, change is 0.1121
elev  6 is  0.6819 inches from elev 0, change is 0.1111
elev  7 is  0.7920 inches from elev 0, change is 0.1101
elev  8 is  0.9010 inches from elev 0, change is 0.1091
elev  9 is  1.0091 inches from elev 0, change is 0.1081
elev 10 is  1.1163 inches from elev 0, change is 0.1071
```

Figure 8.4 - RDSDEPTH.BAS output.

The effect of variations in elevation ranges is more than the expected higher topology. As the number of elevations exceeds the offset, ghosting starts to appear. If the elevation map has steep drop-offs, pixel stretching will take place. This is particularly noticeable in images that have repeating patterns where the next drop in elevation employs the pixel values created from a prior drop in elevation. See the RDS of cones on page 44. The problem is easily solved by avoiding repeating patterns and/or reducing the number of elevations in the map.

Here is a list of criteria that increase the chance of producing a quality stereogram:

- Choose an interesting topic.
- Make sure the object does not have steep drop-offs or repeating patterns.
- Use the highest resolution output device available.
- Don't reduce beyond the resolution of the device when rendering PostScript RDSs, as pixel aliasing results.
- Design for far-eyed viewing to make the images appear larger.
- Don't exceed 80% of your eye spacing for far-eyed RDS offsets.
- Present a zero elevation border adjacent to the sight marks.
- Don't use a perimeter border line, it confuses viewers.
- Normalize the elevations, so the object doesn't float too high above the background.
- Use as many elevations as possible.
- Ensure that the resulting first-glance image has uniform texture and no stretch marks.
- Print so the printer's paper travel is perpendicular to a horizontal viewing line. This minimizes background warps caused by the paper drive mechanisms not maintaining a constant speed. This is also seen on Super VGA screens, due instead to the difficulty of beam alignment of high-resolution computer screens.

Figure 8.5 - TORUS4.PS printed on a 1200 DPI Linotronic printer.

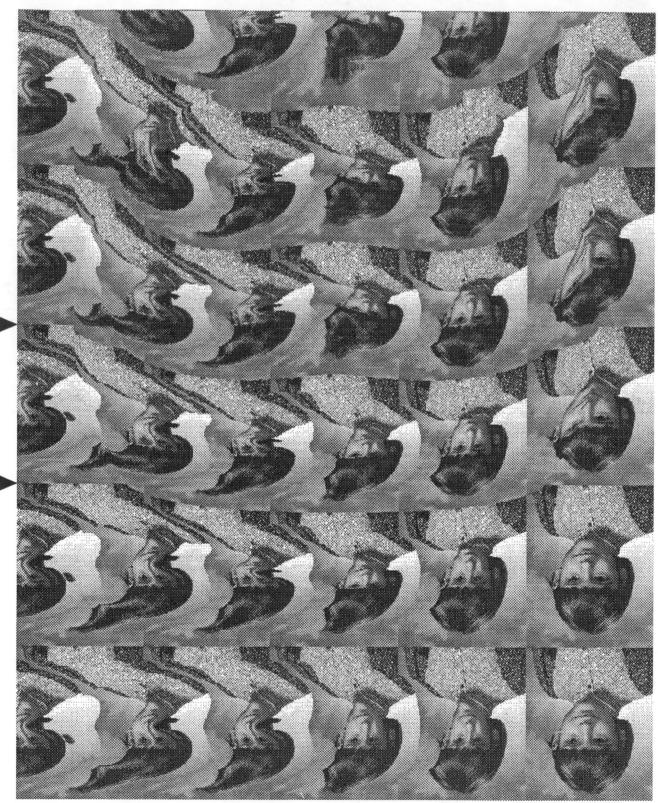

Chapter 9   The Gallery

It is intentional that most of the subjects of the stereograms aren't identified. This was done to keep the uninitiated honest. Would you believe that some people actually try to bluff their way out of not being able to see them? For this reason titles are listed with the gray-scale equivalents in Appendix E.

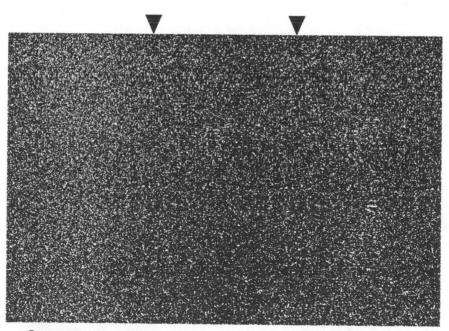

© A. Kinsman 1992

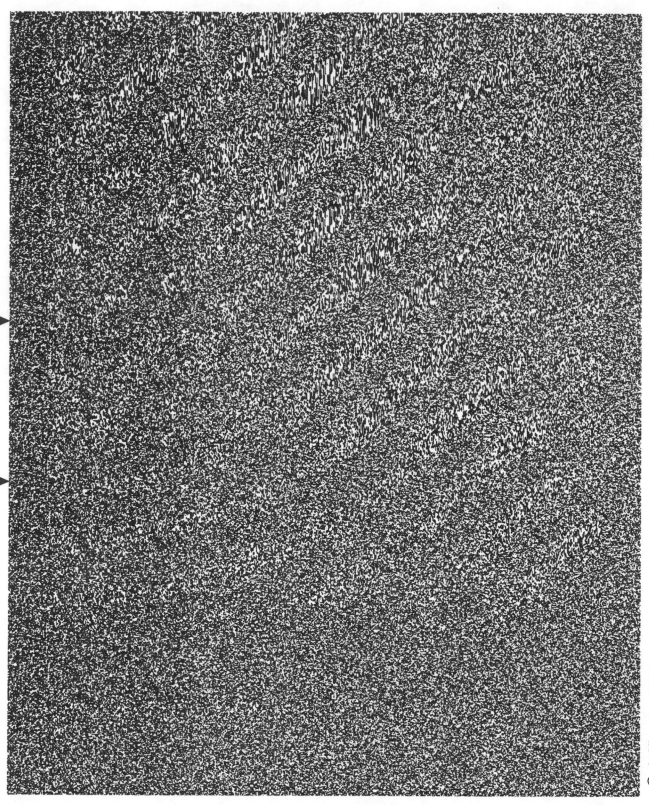

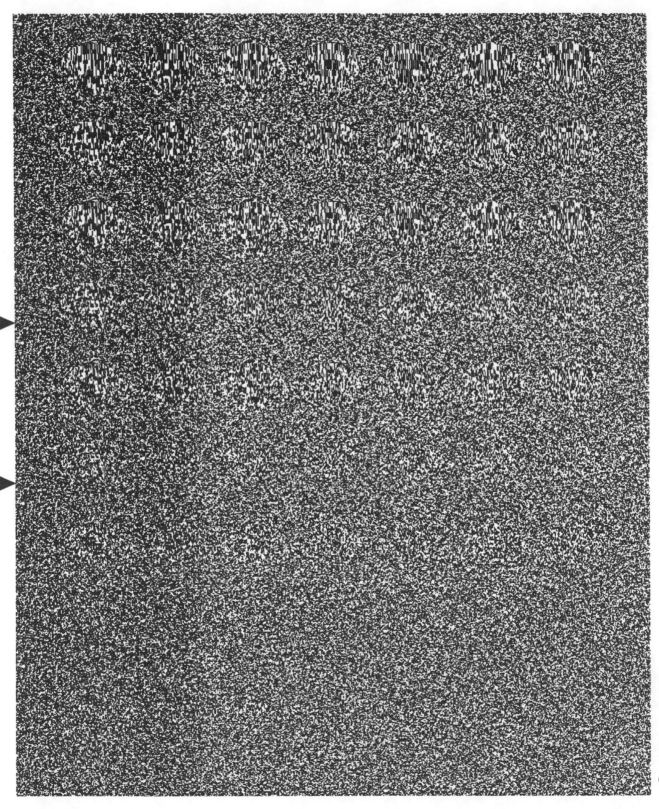

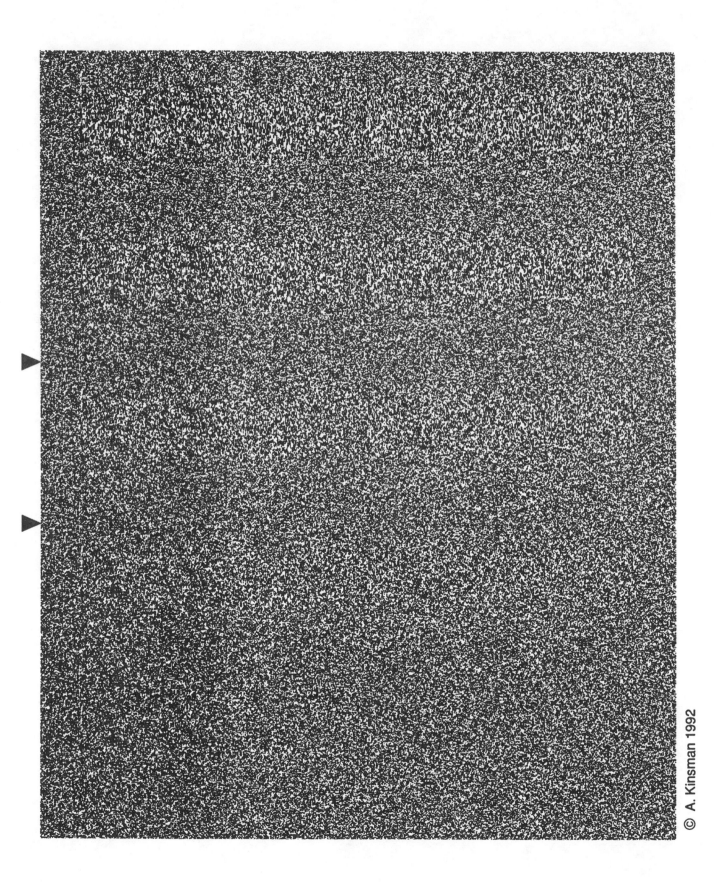

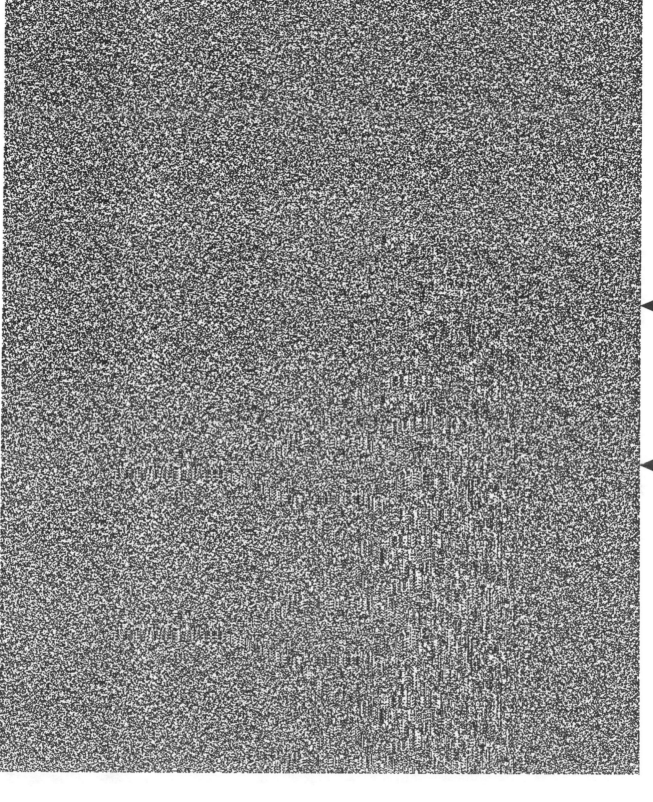

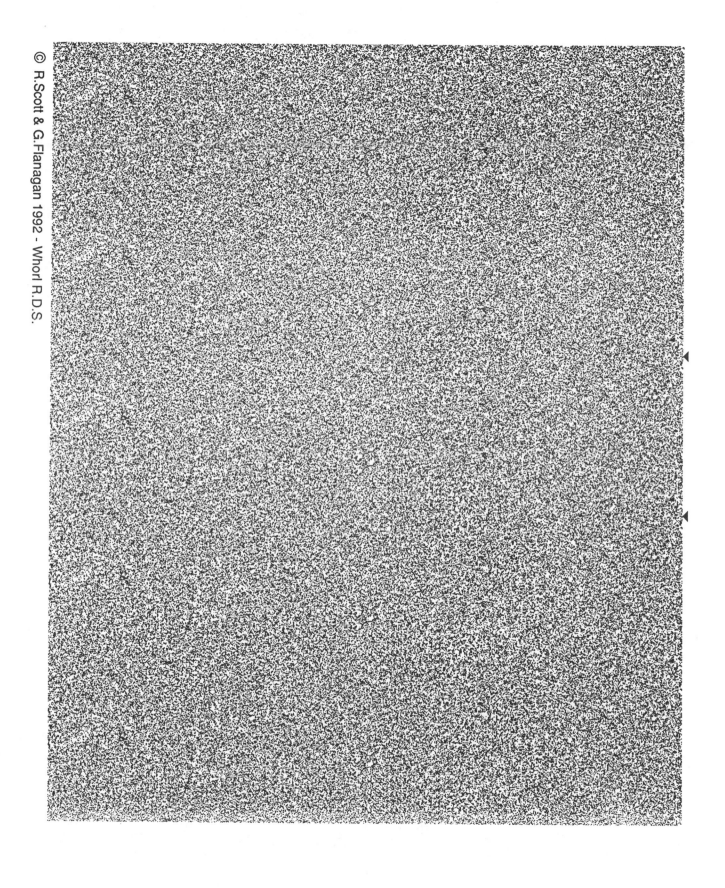

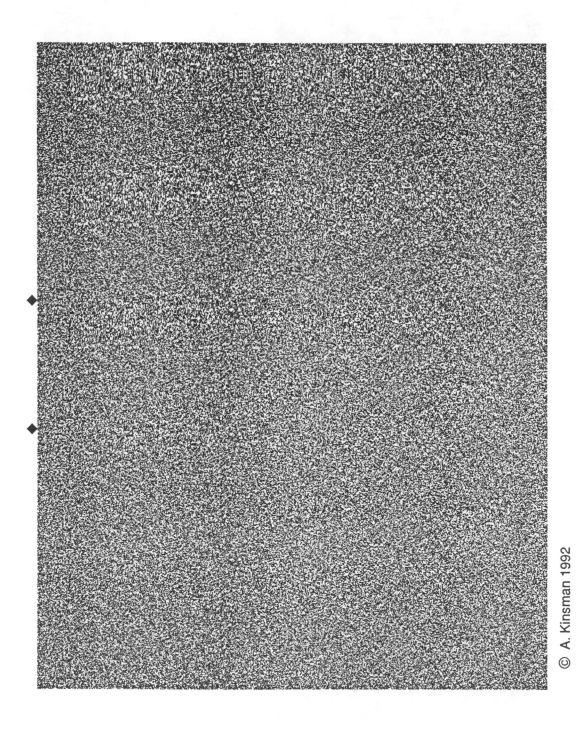

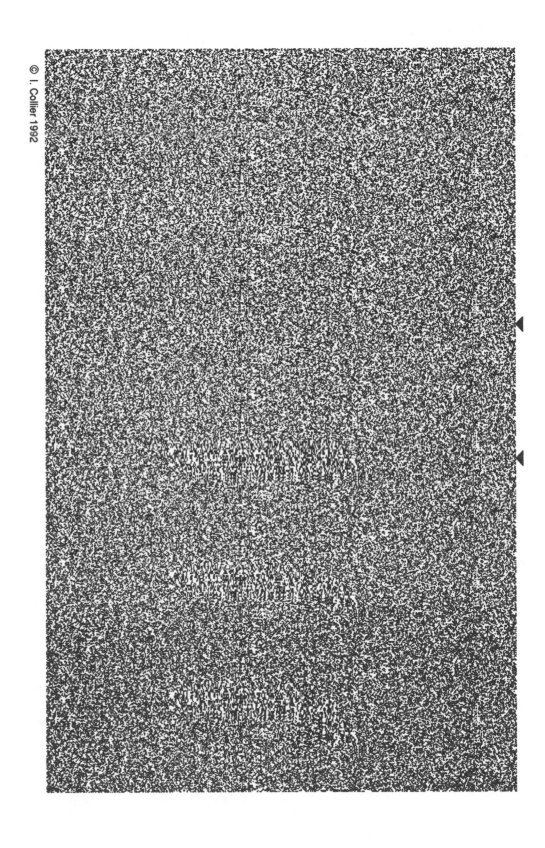

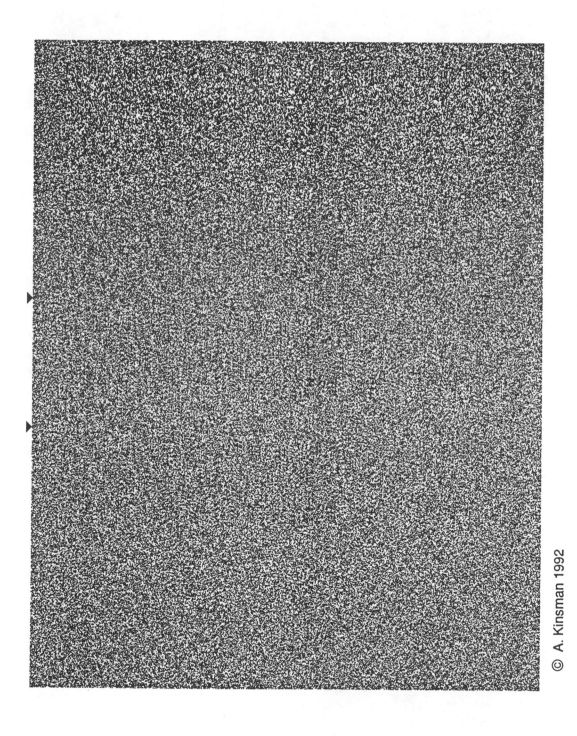

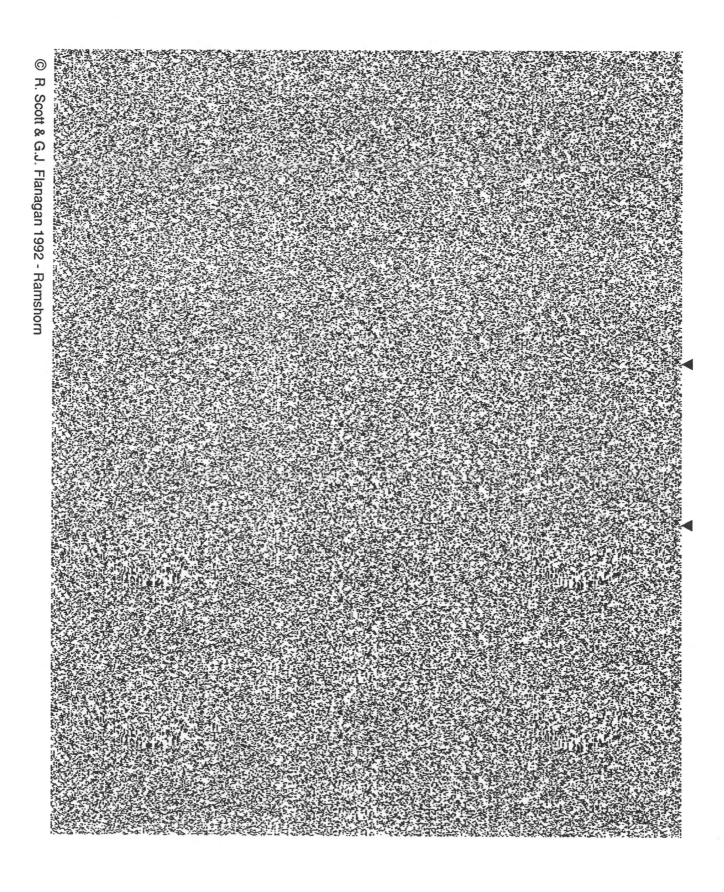

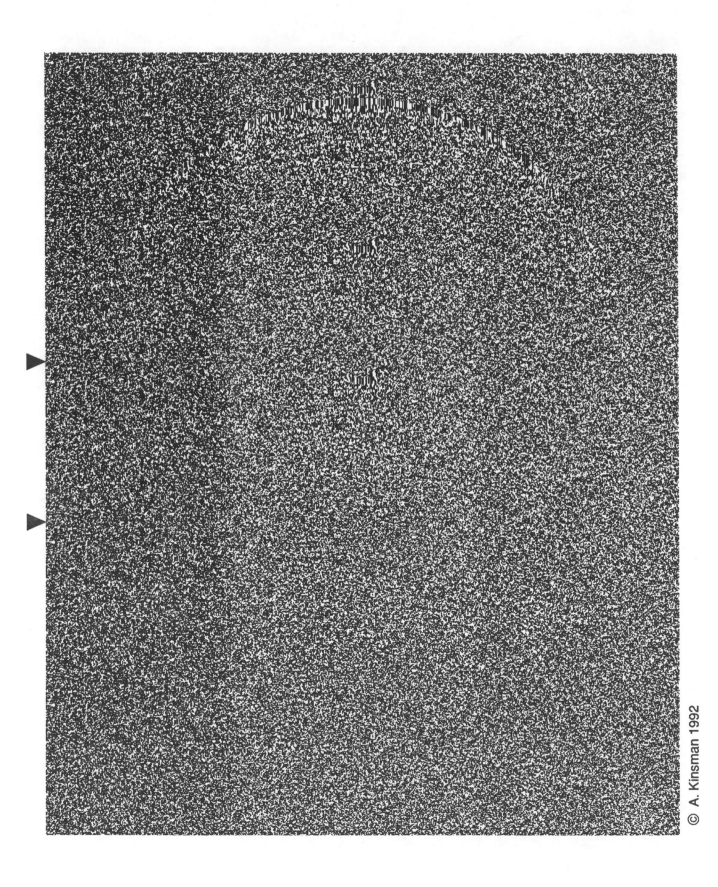

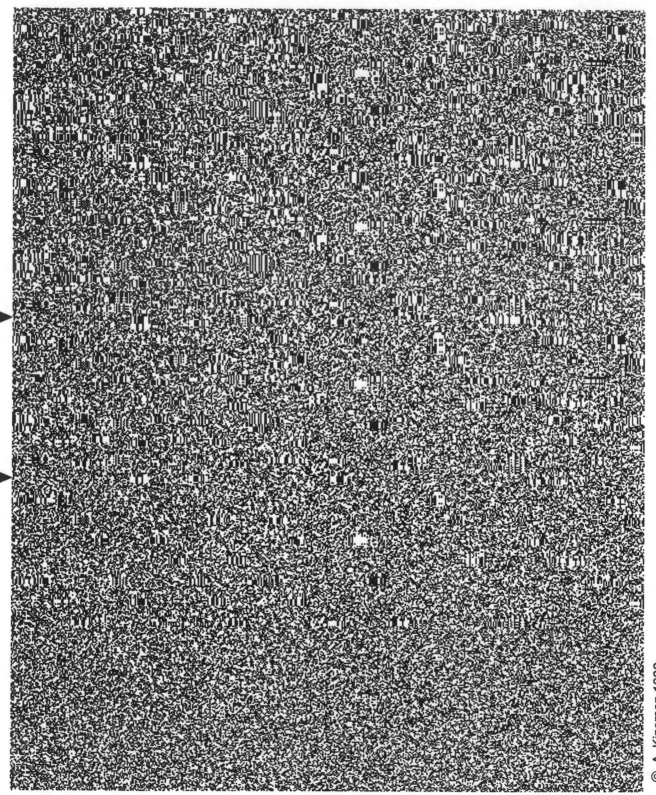

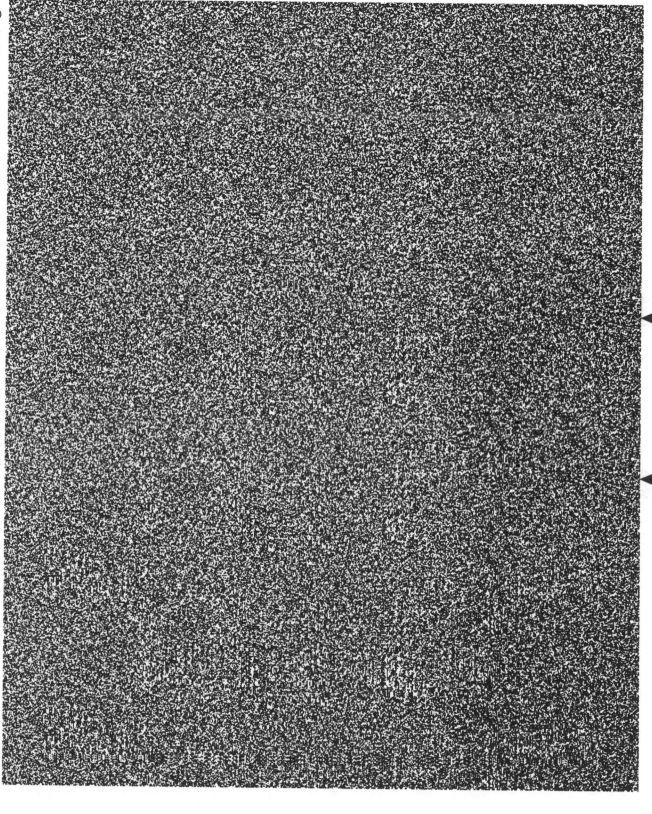

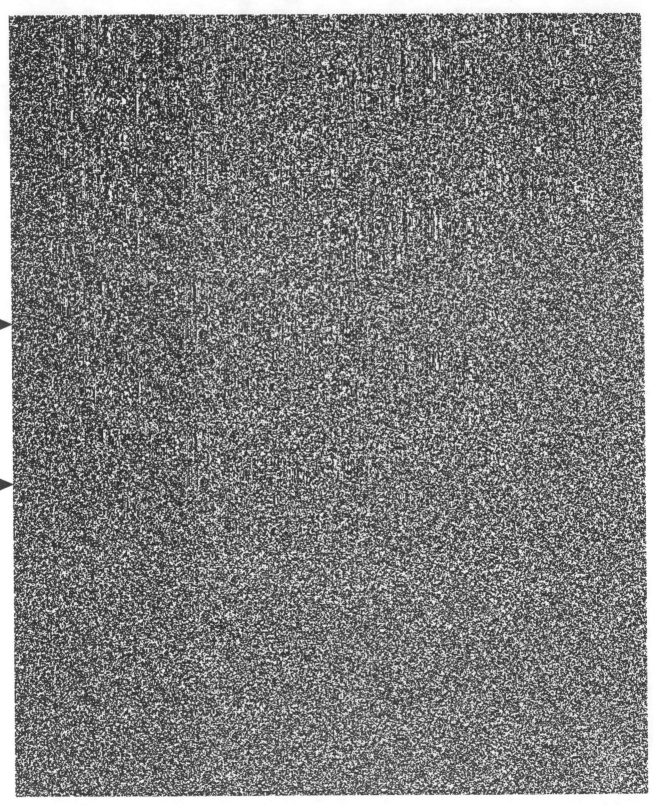

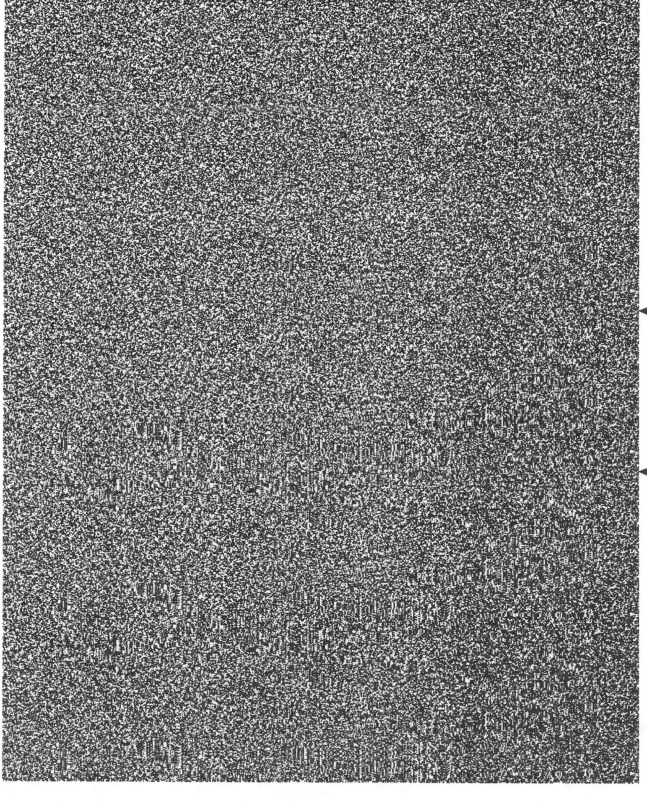

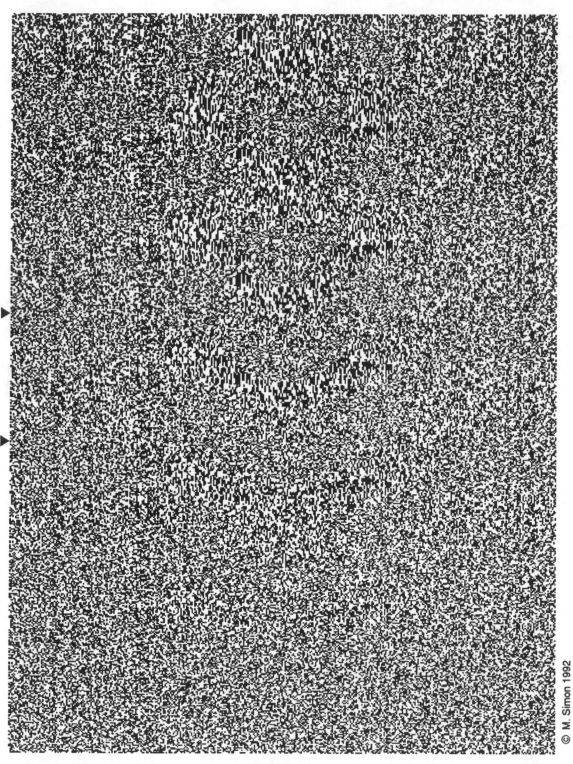

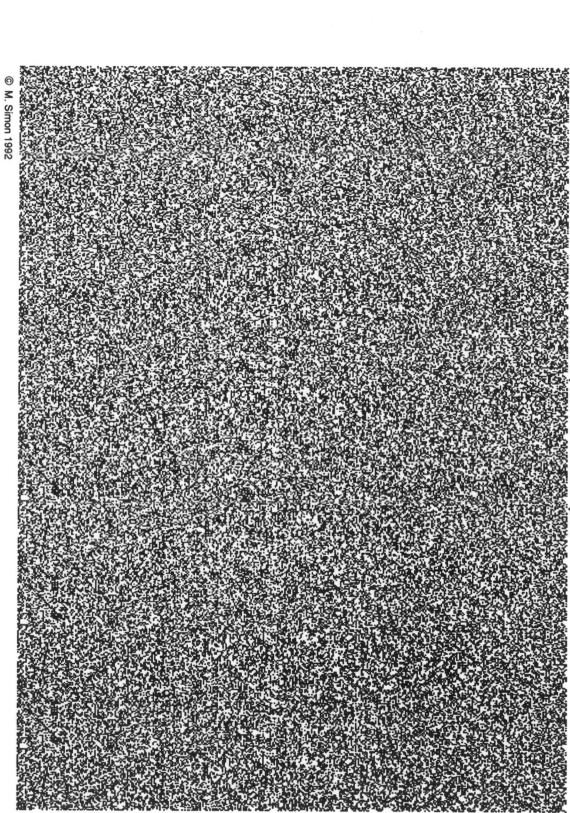

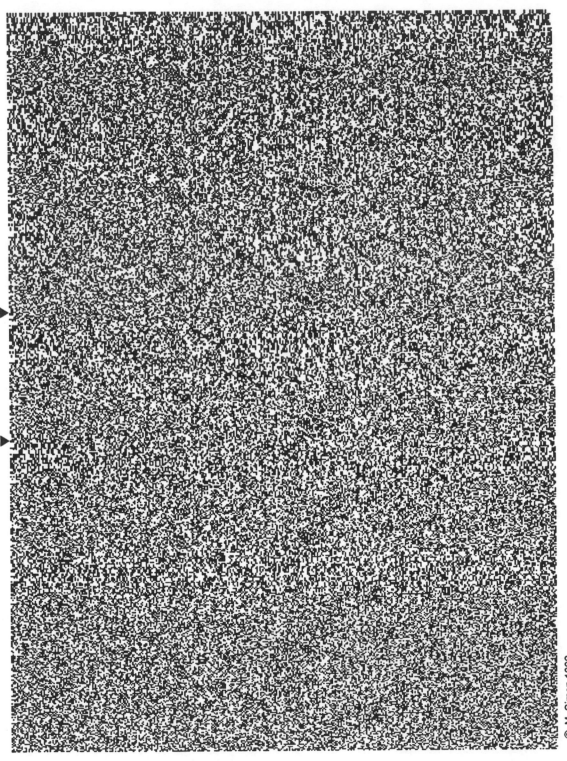

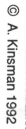

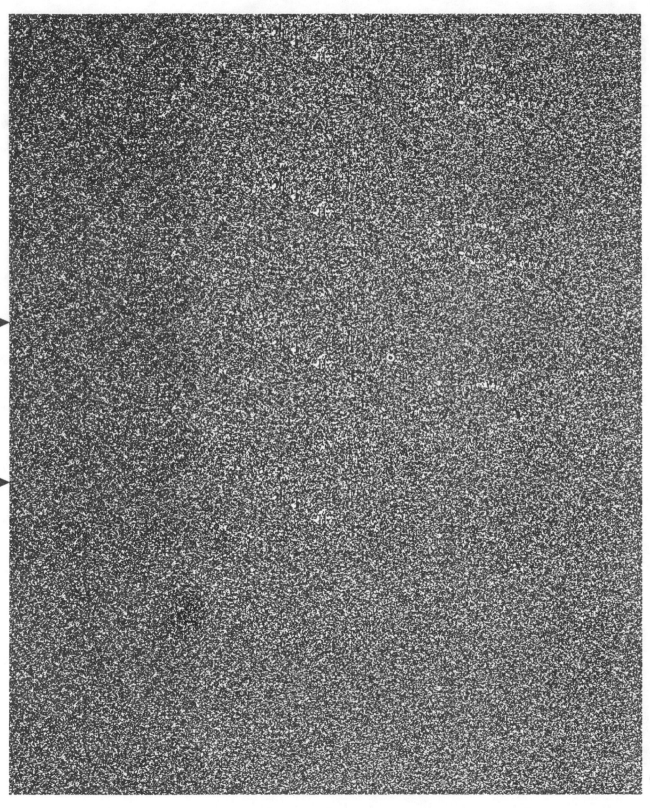

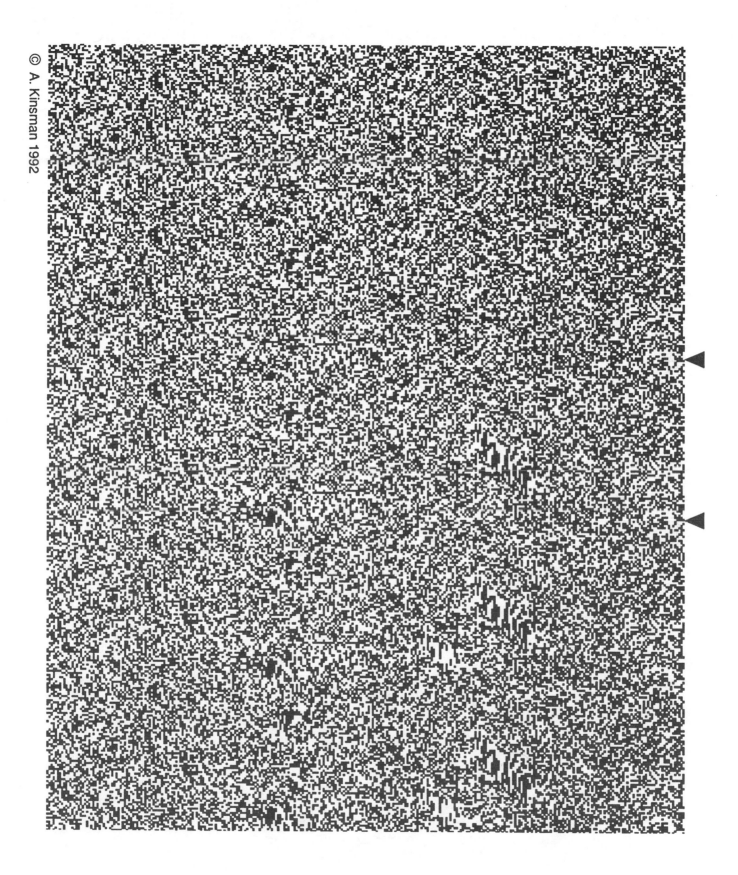

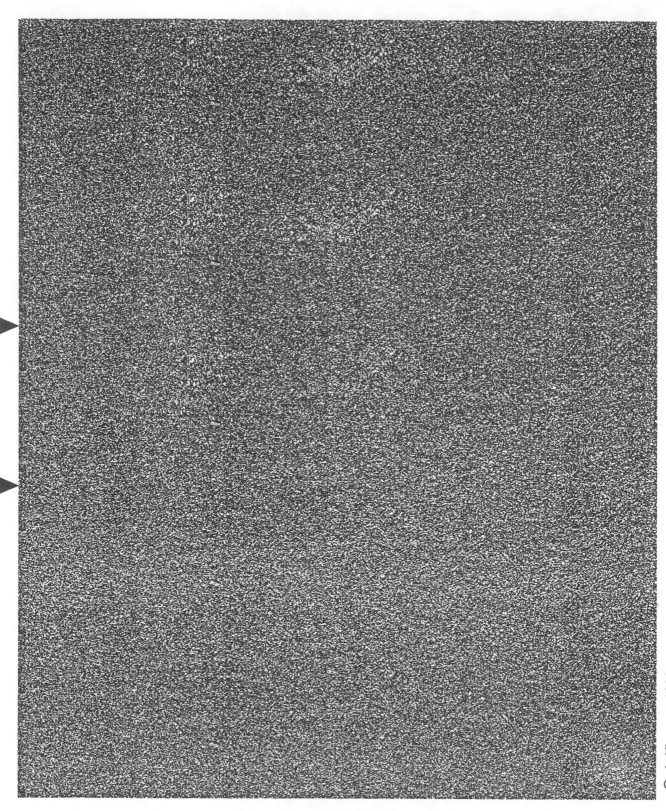

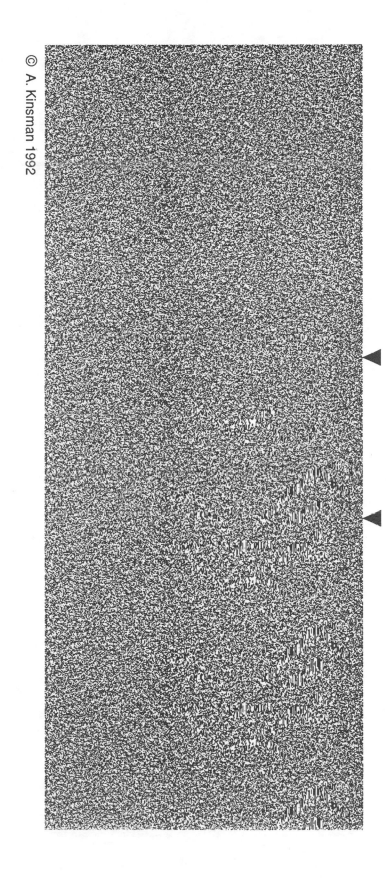

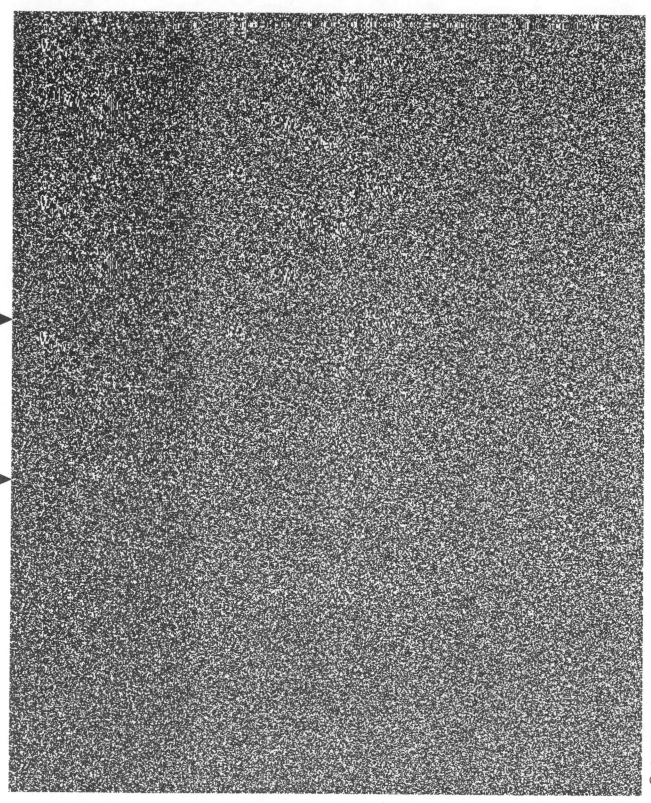

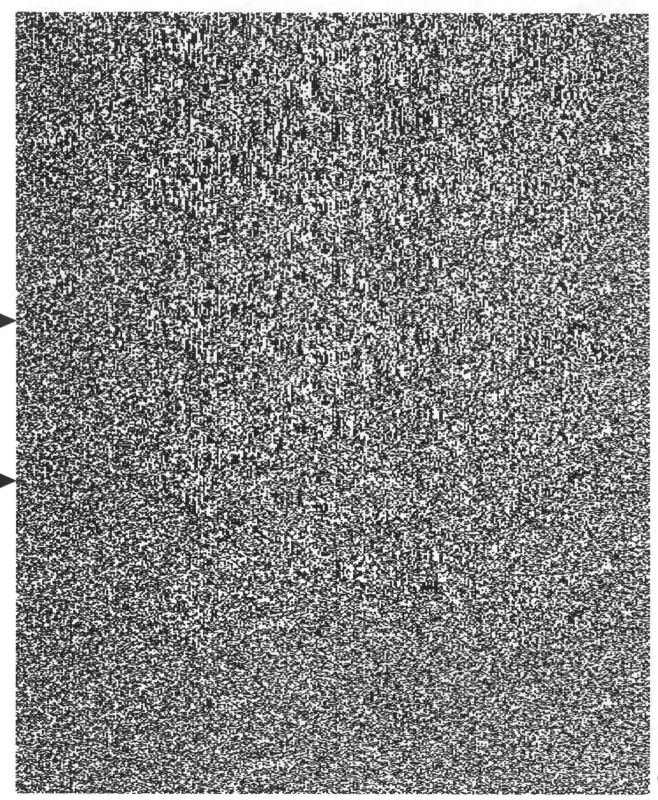

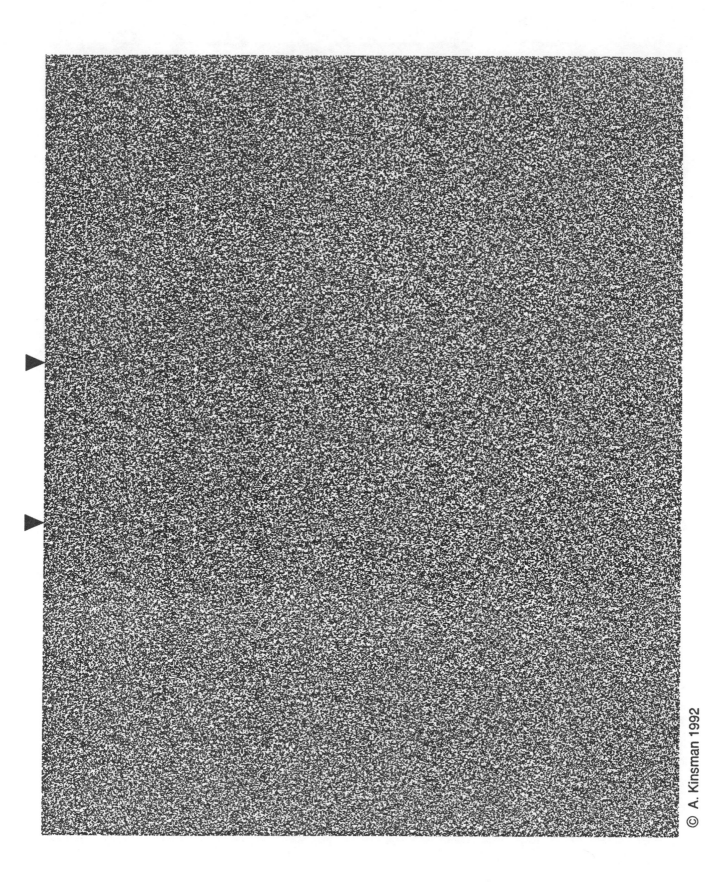

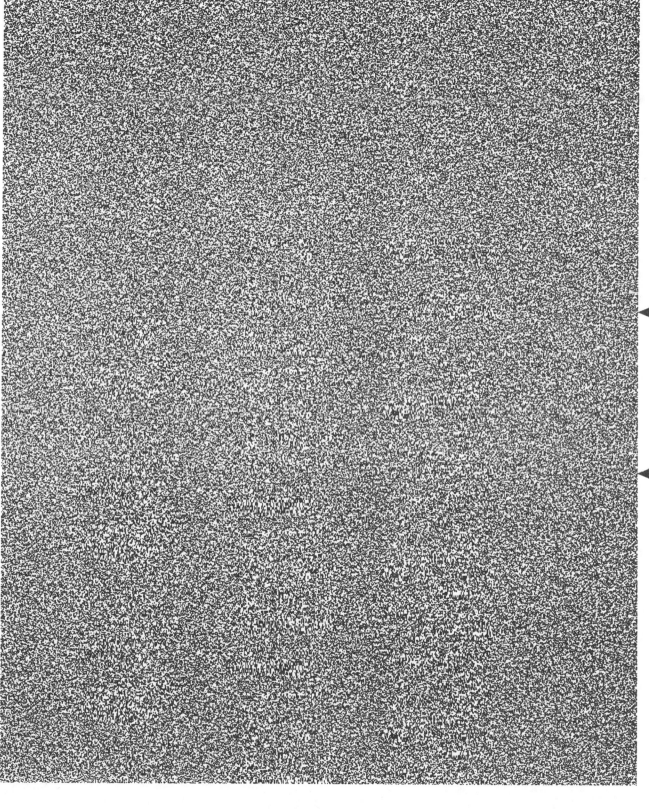

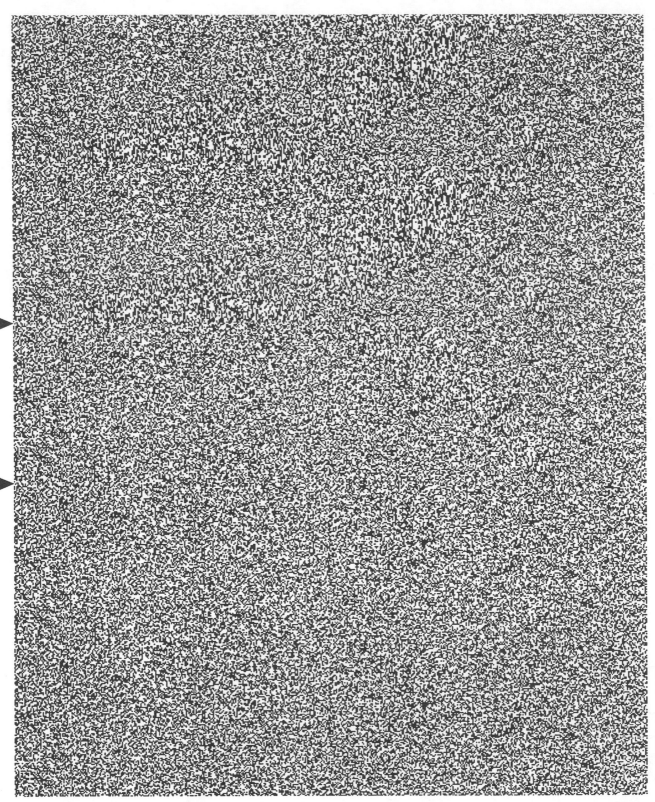

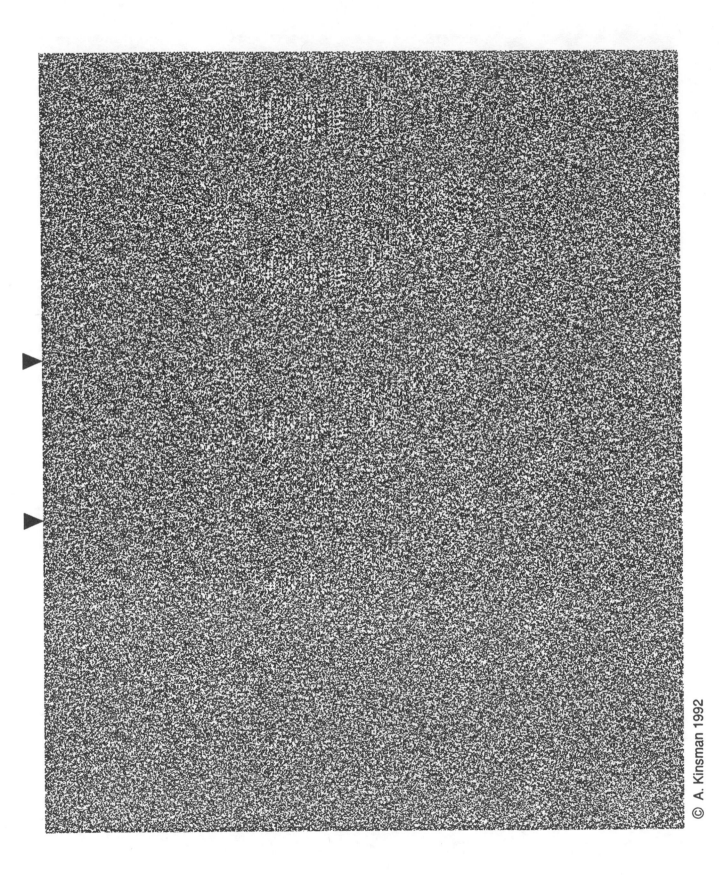

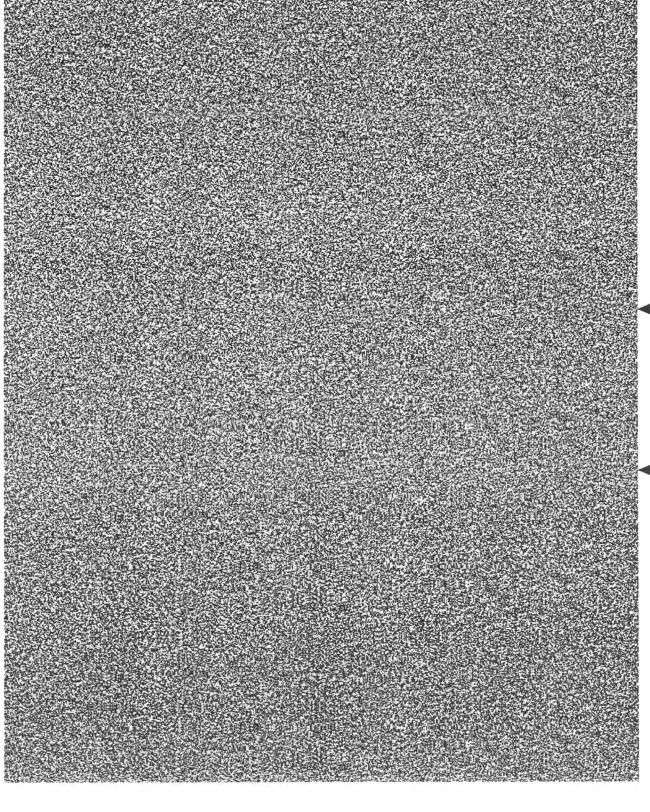

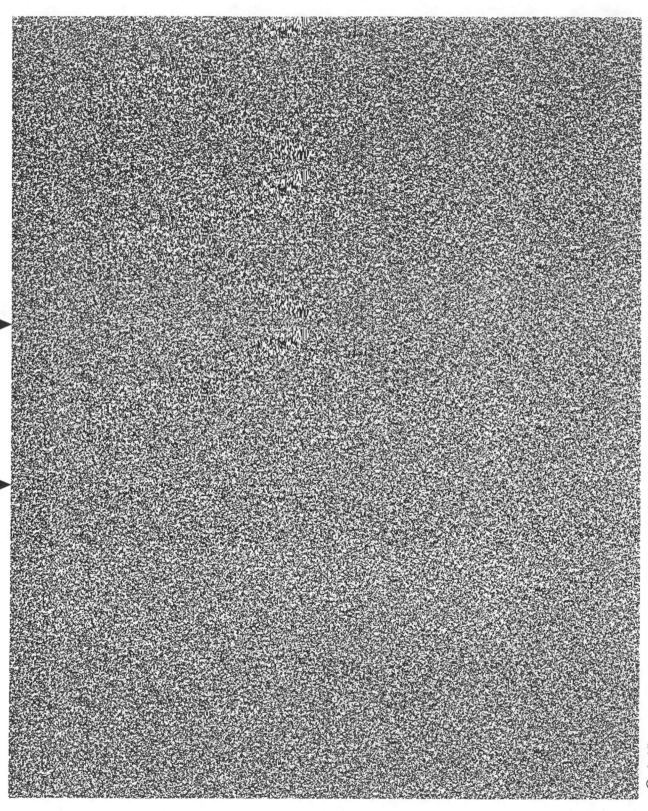

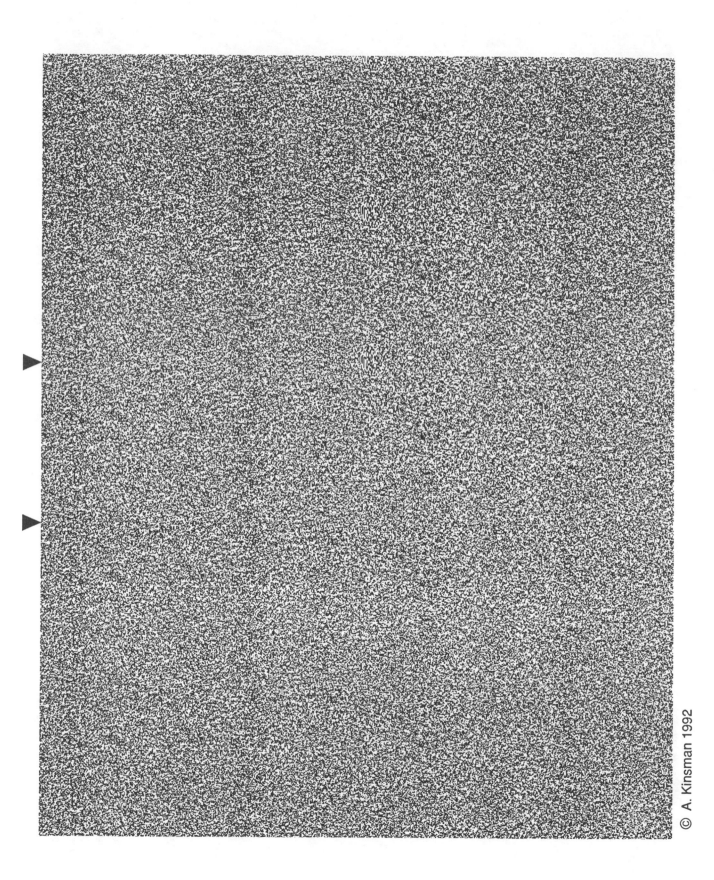

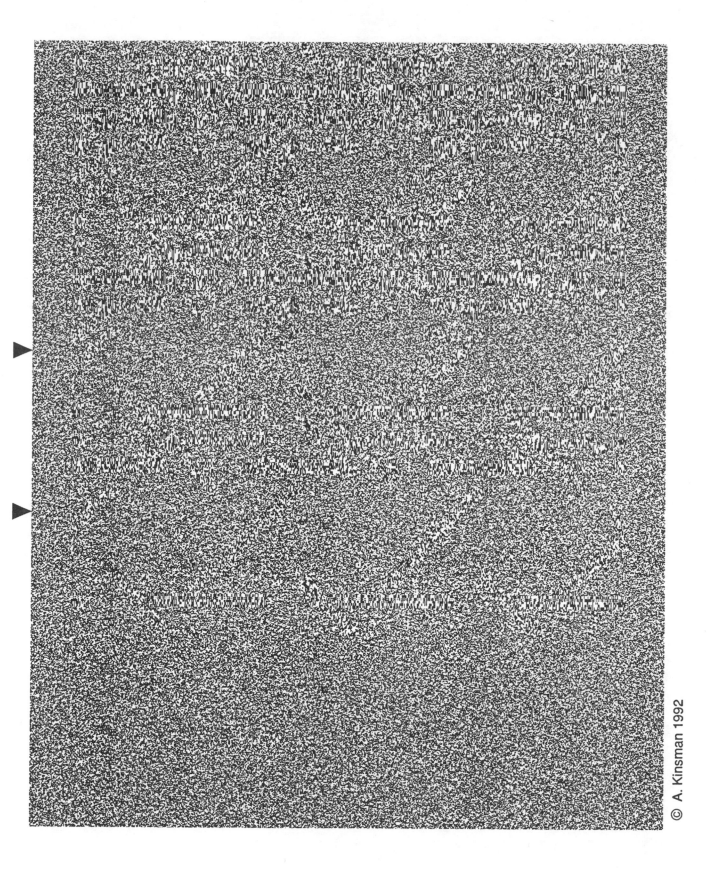

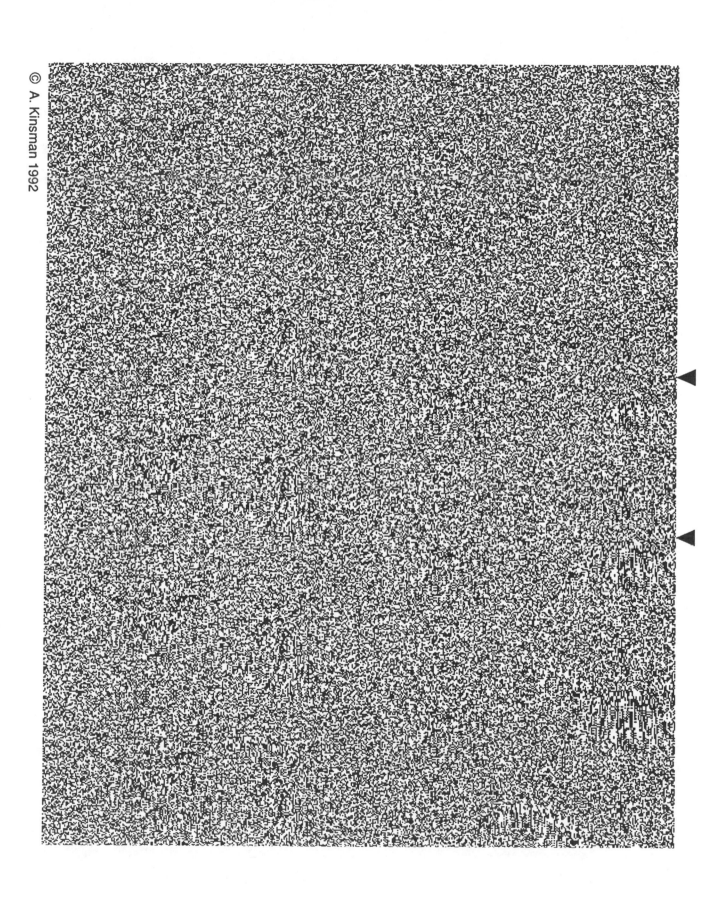

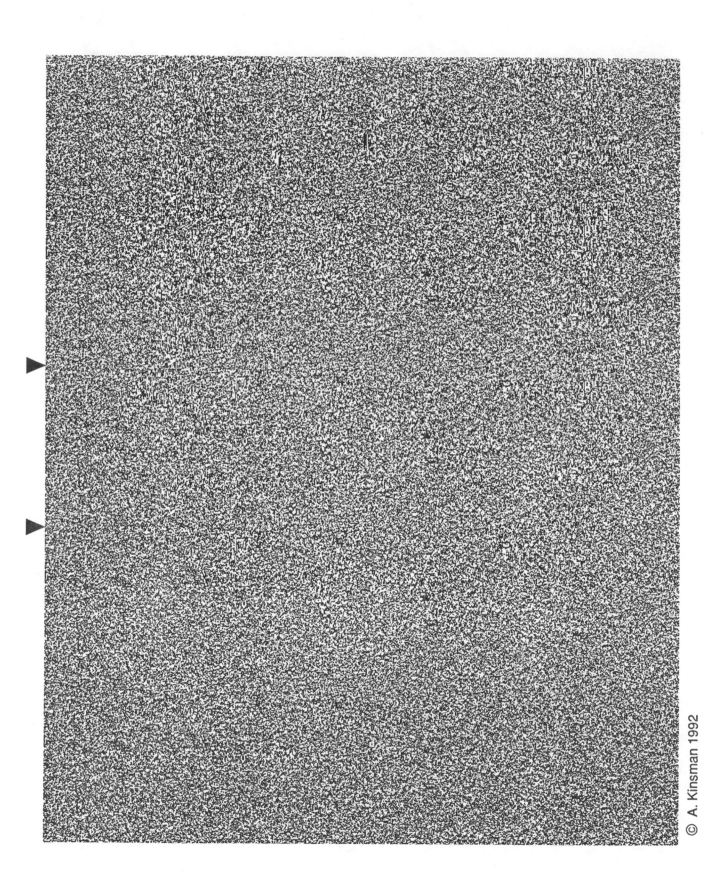

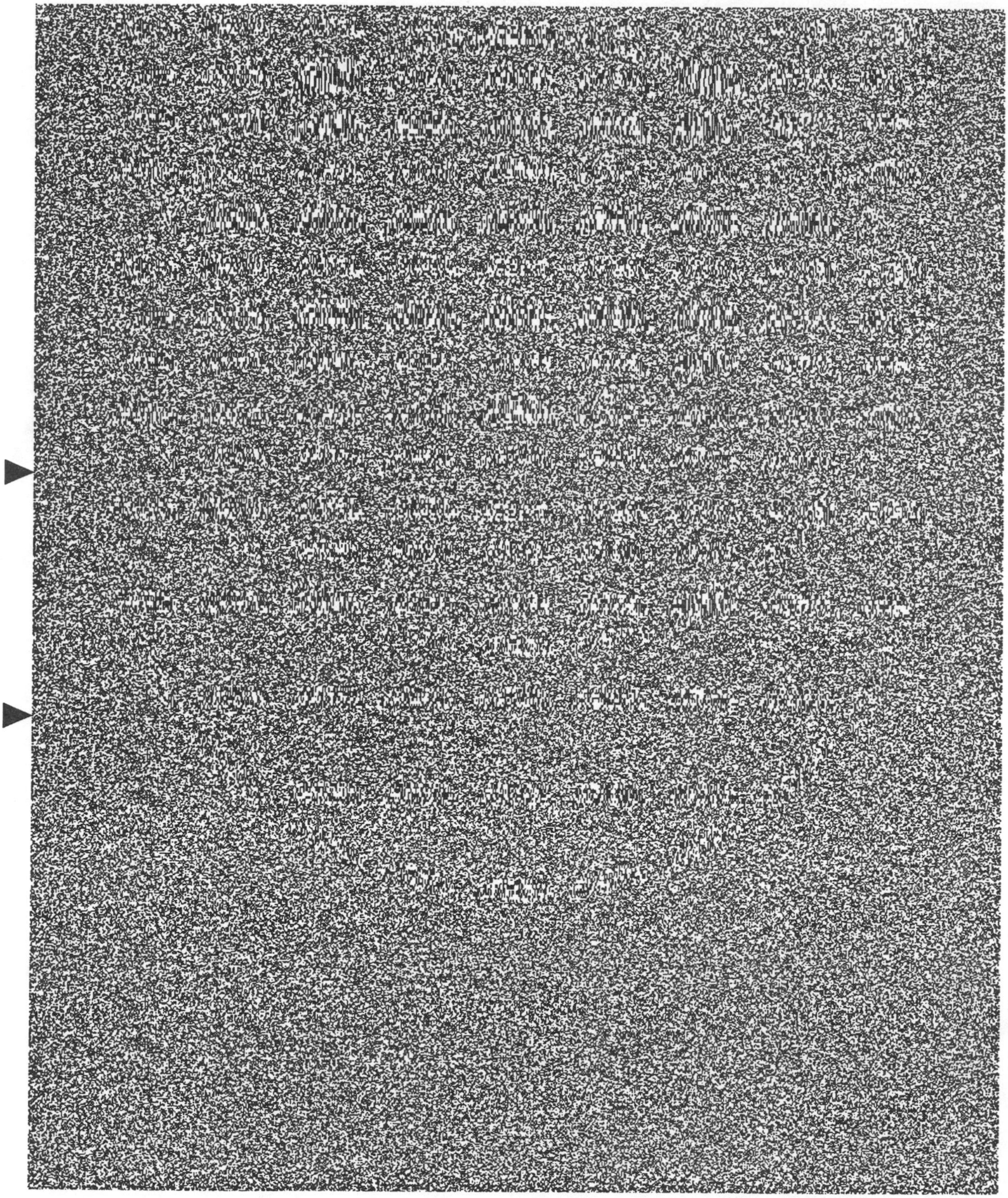

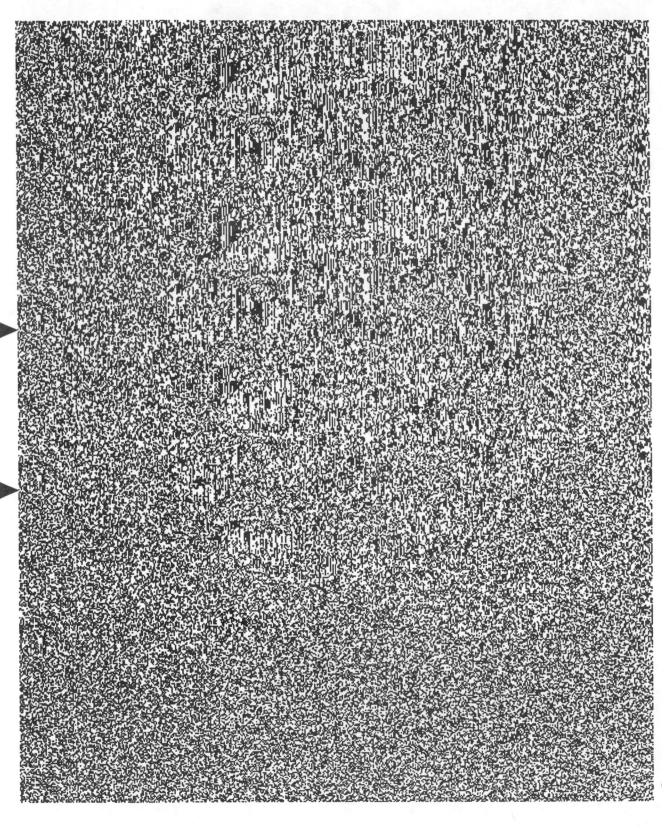

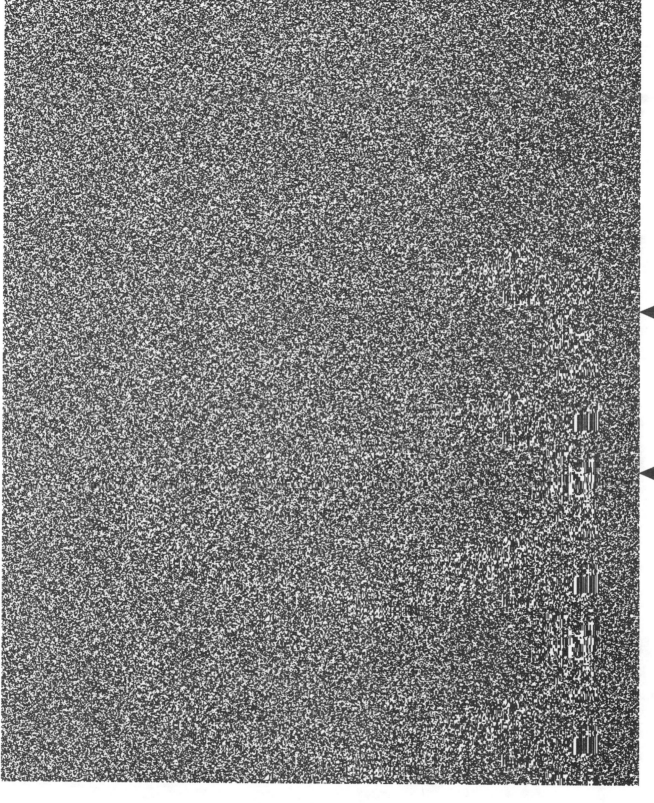

Chapter 10   More Information

For those interested in finding more examples of SIRDS, DIRDS, and random dot anaglyphs, here are some excellent places to begin looking:

> Baylow Productions
> Box 4434
> Long Beach, CA 90804

Mr. Balogh sells a delightful line of anaglyphic post cards. Included with each is a pair of red/green glasses to assist in viewing them. One card is an anaglyph of a random dot stereogram, an excellent method for viewing one if you are unsuccessful with other techniques.

> Reel 3-D Enterprises, Inc.
> P.O. Box 2368
> Culver City, CA 90231

Reel 3-D sells all manner of stereo viewing and related devices. They currently have a wonderful 30-page catalog.

> N.E. Thing Enterprises
> P.O. Box 1827
> Cambridge, MA 02139

N.E. Thing sells random dot stereogram posters and computer software for generating them. They have several brochures available by mail. Products also include calendars and T-shirts.

> Cygnus Graphic
> P.O. Box 32461-D
> Phoenix, Arizona 85064-2461

Cygnus Graphics carries a large selection of anaglyphs, posters, maps, and books all relating to 3-D. Their catalog is available for $1.00 on the North American continent, and $2.00 elsewhere. This is the best collection of 3-D products offered in one place.

> Stereo World
> P.O. Box 14801
> Columbus, OH 43214

Stereo World is the bimonthly publication of National Stereoscopic Association, Inc. The magazine caters to individuals who are serious about viewing, keeping, and collecting stereo photographs. Occasionally random dot stereograms are published. A subscription is automatic when you become a member of the NSA. Annual dues are around $30 U.S.

Appendix A   ASCII Stereogram Programs

A 3.5 inch floppy disk with the source code for the programs listed in this book is available from the publisher for the fee of $10.00 U.S. Please refer to the reverse side of the title page for the publisher's address.

```c
/*
 *      Stereogram.c - produce ASCII stereograms.
 *
 *              "human readable version".
 *
 *      By Greg Alt - Used with permission.
 */

#include <math.h>
#include <stdio.h>
main()
{
    char mask[81],out[81],shift,i;

    /* print line at top to use as a guide for viewing */
    printf("%31sX%15sX\n","","");

    /* read in each mask until end of input */
    while(gets(mask)!=NULL)
    {
        /* create output string one character at a time */
        for(out[80]=shift=i=0;i<80;i++)
        {
            /* character is random if it comes right after a shifted portion */
            /*    or if it is in the first 16 characters */
            if((shift && mask[i-16]!='#') || i<16)
            {
                shift=0;
                /* random character between 'A' and 'Z' */
                out[i]=random()%26+'A';
            }
            else
            {
                /* if the mask character is a '#' then shift */
                shift=(mask[i-16]=='#')?1:0;
                /* copy the character from previous column, shifting if necessary */
                out[i]=out[i-16+shift];
            }
        }
        /* print line */
        puts(out);
    }
}
```

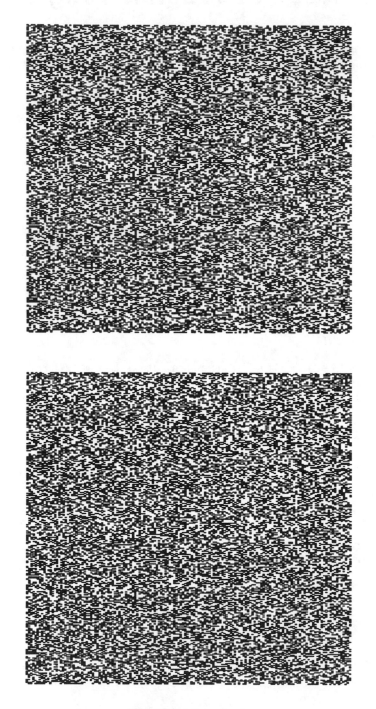

DIRDS.BAS   output.

Appendix B   BASIC Stereogram Programs

```
10 REM Random Dot Stereogram generating program
20 REM RDS.BAS
30 REM By A. Kinsman
40 REM for CGA (640x200) screen
50 SCREEN 2
60 CLS:KEY OFF
70 W=400                    'width of box
80 H=W/2                    'height of box
90 OFFSET= W/4              'separation between sight marks
100 REM fill in left image with random noise
110 REM store data in frame buffer, retrieve with "point" function
120 FOR Y=0 TO H-1
130 't=RND(-.123)    'uncomment to generate random LINE versions
140 FOR X=0 TO OFFSET
150 IF((INT(RND*2))=1 ) THEN PSET(X,Y),1
160 NEXT X
170 NEXT Y
180 REM fill in right side as a function of elevation
190 FOR Y=0 TO H-1
200 FOR X=0 TO W
210 REM use some interesting equation that generates
220 REM an elevation from 0->n, here a CIRCLE
230 REM uncomment the method you prefer to view
240 ELEV = 0
250 REM translate cx and cy axis origin to center of box
260 CX=X/2-H/2    '-h<cx<h
270 CY=Y-H/2      '-h<cy<h
280 IF( SQR((CX^2)+(CY^2)) <45) THEN ELEV=10    'far eyed
290 'IF( SQR((CX^2)+(CY^2)) <45) THEN ELEV=-10 'x-eyed
300 PSET( OFFSET+X,Y), POINT( X+ELEV,Y)
310 NEXT X
320 NEXT Y
330 REM ADD SIGHT MARKS
340 MARK=(W+OFFSET)/2 - OFFSET/2
350 FOR Y=0 TO 5
360 FOR X=0 TO 10
370 PSET( X+MARK         ,Y)
380 PSET( X+MARK+OFFSET,Y)
390 NEXT X
400 NEXT Y
410 IF( INKEY$="" ) GOTO 410     'wait for operator
```

```
 10 REM Dual Image Random Dot Stereogram generating program
 20 REM DIRDS.BAS
 30 REM By A. Kinsman
 40 REM for CGA (640x200) screen
 50 SCREEN 2
 60 CLS:KEY OFF
 70 W=300                     'width of each box
 80 H=W/2                     'height of each box
 90 SEP=640-(W*2)             'separation between boxes
100 REM fill in left image with random noise
110 REM store data in frame buffer, retrieve with "point" function
120 FOR Y=0 TO H
130 't=RND(-.123)    'uncomment to generate random LINE versions
140 FOR X=0 TO W
150 IF((INT(RND*2))=1 ) THEN PSET(X,Y),1
160 NEXT X
170 NEXT Y
180 REM fill in right side as a function of elevation
190 FOR Y=0 TO H
200 FOR X=0 TO W
210 REM use some interesting equation that generates
220 REM an elevation from 0->n, here a CIRCLE
230 REM uncomment the method you prefer to view
240 ELEV = 0
250 REM create cx and cy axis with origin at center of square
260 CX=X/2-H/2       '-h<cx<h now
270 CY=Y-H/2         '-h<cy<h now
280 IF( SQR((CX^2)+(CY^2)) <45) THEN ELEV=-10 'x-eyed
290 'IF( SQR((cx^2)+(cy^2)) <45) THEN ELEV=10 'far-eyed
300 PSET( W+SEP+X,Y), POINT( X+ELEV,Y)
310 'PSET( W+SEP+W-X,Y), POINT( X+ELEV,Y)    'for mirrored DIRDS
320 NEXT X
330 NEXT Y
340 IF( INKEY$="" ) GOTO 340     'wait for operator
```

```
 10 REM Produce a diamond pattern in an elevation map
 20 REM DIAMOND.BAS
 30 REM By A. Kinsman
 40 REM
 50 WIDE  =200               'width of elevation map
 60 HEIGHT=100               'height of elevation map
 70 OPEN "diamond.elv" FOR OUTPUT AS #1
 80 FOR Y = 1 TO HEIGHT
 90 CY = Y-(HEIGHT/2)        'move origin to center
100 FOR X = 1 TO WIDE
110 CX = X-(WIDE/2)          'move origin to center
120 ELEV = 50-( ABS(CX) + ABS(CY)) 'generate some elevation
130 IF ELEV < 0  THEN ELEV=0  'negative clip
140 REM Never use elevation 26 as it is the DOS end of file character
150 IF ELEV = 26 THEN ELEV=25
160 PRINT#1,CHR$(ELEV);        'write as chars to keep file small
170 NEXT X
180 NEXT Y
190 CLOSE #1
```

```
10 REM Random Dot Stereogram generating program - reads elevation map
20 REM RDS-ELV.BAS
30 REM By A. Kinsman
40 REM for CGA (640x200) screen
50 SCREEN 2
60 CLS:KEY OFF
70 INPUT "input file name? ",S$
80 INPUT "what is the elevation map width? ",W
90 INPUT "what is the elevation map height? ",H
100 CLS
110 OPEN S$ FOR INPUT AS #1
120 OFFSET= W/4            'separation between sight marks
130 REM fill in left image with random noise
140 REM store data in frame buffer, retrieve with "point" function
150 FOR Y=0 TO H-1
160 't=RND(-.123)    'uncomment to generate random LINE versions
170 FOR X=0 TO OFFSET
180 IF((INT(RND*2))=1 ) THEN PSET(X,Y),1
190 NEXT X
200 NEXT Y
210 REM fill in right side as a function of elevation
220 REM read in an elevation map from a file
230 REM map has character values from 0->255 in it.
240 FOR Y=0 TO H-1
250 FOR X=0 TO W-1
260 E$=INPUT$(1,#1)       'get elevation file value
270 ELEV = ASC(E$)        'convert back to integer
280 REM rescale all elevation values  elev:255 elev':offset
290 ELEV = (ELEV*OFFSET)/255
300 'elev = -elev            'UNCOMMENT if you view x-eyed
310 PSET( OFFSET+X,Y), POINT( X+ELEV,Y)
320 NEXT X
330 NEXT Y
340 CLOSE #1
350 REM ADD SIGHT MARKS
360 MARK=(W+OFFSET)/2 - OFFSET/2
370 FOR Y=0 TO 5
380 FOR X=0 TO 10
390 PSET( X+MARK          ,Y)
400 PSET( X+MARK+OFFSET,Y)
410 NEXT X
420 NEXT Y
430 IF( INKEY$="" ) GOTO 430     'wait for operator
```

```
 10 REM Translucent Random Dot Stereogram generating program
 20 REM TRDS1.BAS
 30 REM By A. Kinsman
 40 REM pattern is UD
 50 REM              UD
 60 REM for CGA (640x200) screen
 70 SCREEN 2
 80 CLS:KEY OFF
 90 W=400                    'width of box
100 H=W/2                    'height of box
110 OFFSET= W/4              'separation between sight marks
120 REM fill in left image with random noise
130 REM store data in frame buffer, retrieve with "point" function
140 FOR Y=0 TO H-1
150 't=RND(-.123)    'uncomment to generate random LINE versions
160 FOR X=0 TO OFFSET
170 IF((INT(RND*2))=1 ) THEN PSET(X,Y),1
180 NEXT X
190 NEXT Y
200 REM fill in right side as a function of elevation
210 FOR Y=0 TO H-1
220 FOR X=0 TO W STEP 2
230 REM use some interesting equation that generates
240 REM an elevation from 0->n, here a COSINE buldge
250 REM uncomment the method you prefer to view
260 ELEV = 0
270 REM translate cx and cy axis origin to center of box
280 CX=X/2-H/2     '-h<cx<h
290 CY=Y-H/2       '-h<cy<h
300 RADIUS = SQR((CX^2)+(CY^2))
310 REM plots far-eyed flat and upper surfaces
320 IF( RADIUS <90) THEN ELEV=(COS(RADIUS*2*3.1415/180)+1)*5
330 PSET( OFFSET+X,Y),   POINT( X        ,Y)         'flat surface
340 PSET( OFFSET+X+1,Y),POINT( X+1+ELEV,Y)        'cosine bump surface
350 NEXT X
360 NEXT Y
370 REM insert sight marks
380 MARK=(W+OFFSET)/2 - OFFSET/2
390 FOR Y=0 TO 5
400 FOR X=0 TO 10
410 PSET(X+MARK          ,Y)
420 PSET(X+MARK+OFFSET,Y)
430 NEXT X
440 NEXT Y
450 IF( INKEY$="" ) GOTO 450      'wait for operator
```

```
10 REM Translucent Random Dot Stereogram generating program
20 REM TRDS2.BAS
30 REM By A. Kinsman
40 REM pattern is DD
50 REM              UU
60 REM for CGA (640x200) screen
70 SCREEN 2
80 CLS:KEY OFF
90 W=400                    'width of box
100 H=W/2                   'height of box
110 OFFSET= W/4             'separation between sight marks
120 REM fill in left image with random noise
130 REM store data in frame buffer, retrieve with "point" function
140 FOR Y=0 TO H-1
150 't=RND(-.123)    'uncomment to generate random LINE versions
160 FOR X=0 TO OFFSET
170 IF((INT(RND*2))=1 ) THEN PSET(X,Y),1
180 NEXT X
190 NEXT Y
200 REM fill in right side as a function of elevation
210 FOR Y=0 TO H-1
220 FOR X=0 TO W STEP 2
230 REM use some interesting equation that generates
240 REM an elevation from 0->n, here a COSINE buldge
250 REM uncomment the method you prefer to view
260 ELEV = 0
270 REM translate cx and cy axis origin to center of box
280 CX=X/2-H/2    '-h<cx<h
290 CY=Y-H/2      '-h<cy<h
300 RADIUS = SQR((CX^2)+(CY^2))
310 REM plots far-eyed flat and upper surfaces
320 IF( RADIUS <90) THEN ELEV=(COS(RADIUS*2*3.1415/180)+1)*5
330 'ELEV=-ELEV          'UNCOMMENT for x-eyed
340 IF(( Y MOD 2) = 1) COTO 380
350 PSET( OFFSET+X  ,Y),  POINT( X      ,Y)         'flat surface
360 PSET( OFFSET+X+1,Y),  POINT( X+1    ,Y)         'flat surface
370 GOTO 400
380 PSET( OFFSET+X  ,Y),  POINT( X+ELEV  ,Y)        'cosine bump surface
390 PSET( OFFSET+X+1,Y),  POINT( X+1+ELEV,Y)        'flat surface
400 NEXT X
410 NEXT Y
420 REM insert sight marks
430 MARK=(W+OFFSET)/2 - OFFSET/2
440 FOR Y=0 TO 5
450 FOR X=0 TO 10
460 PSET(X+MARK        ,Y)
470 PSET(X+MARK+OFFSET,Y)
480 NEXT X
490 NEXT Y
500 IF( INKEY$="" ) GOTO 500    'wait for operator
```

```
10 REM Translucent Random Dot Stereogram generating program
20 REM TRDS3.BAS
30 REM By A. Kinsman
40 REM pattern is UUU
50 REM             MMM
60 REM             DDD
70 REM for CGA (640x200) screen
80 SCREEN 2
90 CLS:KEY OFF
100 W=400                    'width  of box
110 H=W/2                     'height of box
120 OFFSET=W/4               'separation between sight marks
130 REM fill in left image with random noise
140 REM store data in frame buffer, retrieve with "point" function
150 FOR Y=0 TO H
160 't=RND(-.123)    'uncomment to generate random LINE versions
170 FOR X=0 TO OFFSET
180 IF((INT(RND*2))=1 ) THEN PSET(X,Y),1
190 NEXT X
200 NEXT Y
210 REM fill in right side as a function of elevation
220 FOR Y=0 TO H-1
230 FOR X=0 TO W
240 REM use some interesting equation that generates
250 REM an elevation from 0->n, here a COSINE buldge
260 REM uncomment the method you prefer to view
270 ELEV = 0
280 REM translate cx and cy axis origin to center of box
290 CX=X/2-H/2    '-h<cx<h
300 CY=Y-H/2      '-h<cy<h
310 RADIUS = SQR((CX^2)+(CY^2))
320 REM far-eyed upper,mid,flat coded
330 IF( RADIUS <90) THEN ELEV=(COS(RADIUS*2*3.1415/180)+1)*5
340 'ELEV=-ELEV              'UNCOMMENT for x-eyed
350 IF( Y MOD 3 = 0 ) THEN PSET( OFFSET+X,Y),POINT( X+ELEV+ELEV,Y) 'upper
360 IF( Y MOD 3 = 1 ) THEN PSET( OFFSET+X,Y),POINT( X+ELEV,Y)      'mid
370 IF( Y MOD 3 = 2 ) THEN PSET( OFFSET+X,Y),POINT( X+0,Y)         'flat
380 NEXT X
390 NEXT Y
400 REM insert sight marks
410 MARK=(W+OFFSET)/2 - OFFSET/2
420 FOR Y=0 TO 5
430 FOR X=0 TO 10
440 PSET(X+MARK          ,Y)
450 PSET(X+MARK+OFFSET,Y)
460 NEXT X
470 NEXT Y
480 IF( INKEY$="" ) GOTO 480     'wait for operator
```

```
10 REM Translucent Random Dot Stereogram generating program
20 REM TRDS4.BAS    3 surface ball
30 REM By A. Kinsman        .
40 REM pattern is UUU
50 REM            MMM
60 REM            DDD
70 REM for CGA (640x200) screen
80 SCREEN 2
90 CLS:KEY OFF
100 W=400                   'width of box
110 H=W/2                   'height of box
120 OFFSET=W/4              'separation between sight marks
130 REM fill in left image with random noise
140 REM store data in frame buffer, retrieve with "point" function
150 FOR Y=0 TO H-1
160 't=RND(-.123)    'uncomment to generate random LINE versions
170 FOR X=0 TO OFFSET
180 IF((INT(RND*2))=1 ) THEN PSET(X,Y),1
190 NEXT X
200 NEXT Y
210 REM fill in right side as a function of elevation
220 FOR Y=0 TO H-1
230 FOR X=0 TO W
240 REM use some interesting equation that generates
250 REM an elevation from 0->n, here a COSINE buldge
260 REM uncomment the method you prefer to view
270 ELEV = 0
280 REM translate cx and cy axis origin to center of box
290 CX=X/2-H/2     '-h<cx<h
300 CY=Y-H/2       '-h<cy<h
310 RADIUS = SQR((CX^2)+(CY^2))
320 REM far-eyed or x-eyed viewing
330 IF( RADIUS <90) THEN ELEV=90^2-(CX^2+CY^2)
340 ELEV=SQR(ELEV)/5
350 IF( Y MOD 3 = 0 ) THEN PSET( OFFSET+X,Y),POINT( X+ELEV,Y)
360 IF( Y MOD 3 = 1 ) THEN PSET( OFFSET+X,Y),POINT( X+0,Y)
370 IF( Y MOD 3 = 2 ) THEN PSET( OFFSET+X,Y),POINT( X-ELEV,Y)
380 NEXT X
390 NEXT Y
400 REM insert sight marks
410 MARK=(W+OFFSET)/2 -OFFSET/2
420 FOR Y=0 TO 5
430 FOR X=0 TO 10
440 PSET( X+MARK         ,Y)
450 PSET( X+MARK+OFFSET,Y)
460 NEXT X
470 NEXT Y
480 IF( INKEY$="" ) GOTO 480    'wait for operator
```

```
 10 REM compute apparent far-eyed RDS depth values
 20 REM rdsdepth.bas
 30 REM by A. Kinsman 8/18/92
 40 REM
 50 ESET=2.25                'eye spacing for viewer (inches)
 60 VD=24                    'viewing distance from RDS surface (inches)
 70 EPI=100                  'pixels per inch
 80 OFFP=10                  'offset in pixels
 90 EYEP=INT(ESET*EPI)       'offset between eyes in pixels
100 DBASE=VD*OFFP/(EYEP-OFFP) 'depth into screen of base (elev=0)
110 PRINT "base elevation is ";DBASE;" inches behind page"
120 LAST = DBASE
130 FOR I=0 TO OFFP
140 DI=VD*(OFFP-I)/(EYEP-OFFP+I)   'depth into screen of this level
150 DELTA=DBASE-DI
160 PRINT USING "elev ###";I;
170 PRINT USING " is ##.####";DELTA;
180 PRINT " inches from elev 0";
190 PRINT USING ", change is #.####";LAST-DI
200 LAST=DI
210 NEXT I
```

```
 10 REM compute apparent x-eyed RDS depth values
 20 REM rdsdepth2.bas
 30 REM by A. Kinsman 8/18/92
 40 REM
 50 ESET=2.25                'eye spacing for viewer (inches)
 60 VD=24                    'viewing distance from RDS surface (inches)
 70 EPI=100                  'pixels per inch
 80 OFFP=10                  'offset in pixels
 90 EYEP=INT(ESET*EPI)       'offset between eyes in pixels
100 DBASE=VD*OFFP/(EYEP+OFFP)  'dist out of screen for base (elev=0)
110 PRINT "base elevation is ";DBASE;" inches in front of page"
120 LAST = DBASE
130 FOR I=0 TO OFFP
140 DI=VD*(OFFP-I)/(EYEP+OFFP-I)   'dist out of screen of this level
150 DELTA=DBASE-DI
160 PRINT USING "elev ###";I;
170 PRINT USING " is ##.####";DELTA;
180 PRINT " inches from elev 0";
190 PRINT USING ", change is #.####";LAST-DI
200 LAST=DI
210 NEXT I
```

Appendix C   "C" Stereogram Programs

```
/*
 *   randot.c - generate a random dot stereogram from a
 *       collection of elevation data points.
 *
 *   By   A. Kinsman 7/13/92
 *
 *       The elevation data is in values from 0-255.
 *       These values are scaled to max-elev read from the
 *           the command line. max-elev of zero means the
 *           program will select an appropriate scale value
 *           computed from the width.
 *
 * usage:   randot x-dim y-dim max-elev <elevation_data >outputfile
 */
#include <stdio.h>

#define MAXXDIM 4000

main(argc, argv)
int   argc;
char *argv[];
{
  int  i,width,height,max_elev,offset,elev,lines;
  unsigned char elev_vals[MAXXDIM];
  unsigned char one_line [MAXXDIM*2];

  if (argc != 4)
  {
        printf("usage is    randot x-dim y-dim max-elev <elevation_data
>outputfile\n");
        exit(1);
  }

  width   = (int) strtol(argv[1], (char **) NULL, 0);
  height  = (int) strtol(argv[2], (char **) NULL, 0);
  max_elev= (int) strtol(argv[3], (char **) NULL, 0);

  /* if no max elevation defined, compute a good one */
  if( max_elev == 0 )
      max_elev = width/20;

  /* compute a reasonable offset for image */
  offset = width/4;
  if( offset < max_elev ) offset = max_elev;


  lines=0;
  while( !feof(stdin) )
  {
    /* read in a line of elevations */
    for(i=0;i<width;i++)
       elev_vals[i]=getchar();

    for(i=0;(i<width+offset)&&(!feof(stdin));i++)
    {
```

```
            if( i<offset )    /* filling in noise field at left edge */
            {
             one_line[i]=random() % 2;
             if(one_line[i]) one_line[i]=255;

            }
            else            /* working with elevation map values */
            {
                if( i < width+offset )
                {
                    elev=elev_vals[i-offset];
                    /* elev=255-elev;          /* invert altitude */
                    elev=(elev*max_elev)/255; /* scale it */
                }
                else elev=0;  /* off right edge of map */
                one_line[i]=one_line[(i-offset)+elev];
            }
            putchar(one_line[i]);
        }
        lines++;
    }
    lines--;  /* adjustment */

    fprintf(stderr,"resulting file is %d wide and %d lines
long\n",width+offset,lines);
}



/*
 * analyze.c
 *       - check elevation maps, by displaying distribution.
 *
 *   by A. Kinsman
 *
 */

#define MAXZ 256
main()
{
  int val,i;
  unsigned int heights[MAXZ];

            /* clear height array */
  for(i=0;i<MAXZ;i++)
      heights[i]=0;

            /* add to count for each height found */
  while( (val=getchar()) != -1 )
      heights[val]++;

            /* display them scrolling, highest to lowest */
  for(i=MAXZ-1;i>=0;i--)
      printf("height[%03d] used %d times\n",i,heights[i]);
  printf("\n");
}
```

```
/*
 *  bin2hex.c
 *      - convert a binary file to ASCII hex characters.
 *      - put no more than 64 characters on a line.
 *
 *  by A. Kinsman
 */
main()
{
  int val,i;

  i=0;

  while( (val=getchar()) != -1 )
  {
        printf("%2.2x",val);
        i++;
        if( (i % 32) == 0 )
                printf("\n");
  }
  printf("\n");
}


/*
 *  color2bw.c
 *      - pitch 2 out of three bytes, keeping green values only.
 *
 *  by A. Kinsman
 */
main()
{
  int val;

  while( (val=getchar()) != -1 )
  {
      val=getchar();
      putchar( val );
      val=getchar();
  }
}
```

```
/*
 * cutout.c
 *       - cut a chunk out of the file at the specific area of interest.
 *       - width and height are original file size
 *       - loc-x and loc-y are coordinates of upper left corner of area
 *           of interest.
 *       - x-size and y-size are resulting file size.
 *
 *   by A. Kinsman
 */
#include <stdio.h>

main(argc, argv)
int  argc;
char *argv[];
{
   int val,i,j,width,height,locx,locy,xsize,ysize;

   if (argc != 7)
   {
        printf("usage is   cutout width height loc-x loc-y x-size y-size
<inputfile >outputfile\n");
        exit(1);
   }

  width = (int) strtol(*++argv, (char **) NULL, 0);
  height= (int) strtol(*++argv, (char **) NULL, 0);
  locx  = (int) strtol(*++argv, (char **) NULL, 0);
  locy  = (int) strtol(*++argv, (char **) NULL, 0);
  xsize = (int) strtol(*++argv, (char **) NULL, 0);
  ysize = (int) strtol(*++argv, (char **) NULL, 0);


   /* read the whole image, use that in the area of interest */
   for(i=0;i<height;i++)
       for(j=0;j<width;j++)
       {
            val=getchar();

            /* in AOI? */
            if( (i >= locy) &&
                (i <  locy+ysize ) &&
                (j >= locx) &&
                (j <  locx+xsize )     )
                  putchar(val);
       }
}
```

```
/*
 * everyother.c
 *    - discard every other pixel in an image file.
 *
 *      Since this works in both directions, the image will have
 *      1/4 the original area.
 *
 * by A. Kinsman
 *
 *    KDKD...
 *    DDDD....
 *    KDKD.....        K=keep, D=discarded
 */
#include <stdio.h>

main(argc, argv)
int  argc;
char *argv[];
{
  int val,i,j,xsize,ysize;

  if (argc != 3)
  {
       printf("usage is   everyother x-size y-size <inputfile >outputfile\n");
       exit(1);
  }

  xsize= (int) strtol(*++argv, (char **) NULL, 0);
  ysize= (int) strtol(*++argv, (char **) NULL, 0);

  /* process entire image */
  for(i=0;i<ysize;i++)
      for(j=0;j<xsize;j++)
      {
            val=getchar();
            if( (i % 2)== 0 )
            {
                  if( (j%2)==0)
                  {
                      printf("%c",val);
                  }
            }
      }
}
```

```c
/*
 * gray2bit.c
 *  - convert a grayscale (8bit) image that contains
 *      only black and white pixels into a bitmap, making it much
 *      more efficient to store.
 *
 *   by A. Kinsman
 */
#include <stdio.h>

main( argc, argv)
int     argc;
char    *argv[];
{
  int i,val,width,sum,count;

  if (argc != 2)
  {
          printf("usage is   gray2bit image-width <inputfile >outputfile\n");
          exit(1);
  }

  width = (int) strtol(*++argv, (char **) NULL, 0);

  val=getchar();
  while( !feof(stdin) )
  {
      sum = 0;
      count = 0;

          /* process a line */
      for(i=0;i<width;i++)
      {
          if( count < 8 )
          {
                  sum=(sum<<1)|(val?1:0);
                  count++;
          }
          else
          {
                  putchar(sum);
                  sum=val?1:0;
                  count = 1;
          }
          val=getchar();
      }
          /* eol residual dump */
      if( count > 0 )
      {
          while( count < 8 )
          {
                  sum=(sum<<1);
                  count++;
          }
          putchar(sum);
      }
  }
}
```

```
/*
 * replicate.c
 *      - pixel replicate to double the size of an image.
 *
 *      Since this works in both directions, the image will have
 *      4 times the original area.
 *
 * by A. Kinsman
 *
 */
#include <stdio.h>

#define MAXWIDTH  4096

static unsigned char one_line[MAXWIDTH];

main(argc, argv)
int  argc;
char *argv[];
{
  int val,i,j,xsize,ysize;

  if (argc != 3)
  {
        printf("usage is    replicate x-size y-size <inputfile >outputfile\n");
        exit(1);
  }

  xsize= (int) strtol(*++argv, (char **) NULL, 0);
  ysize= (int) strtol(*++argv, (char **) NULL, 0);


  if( xsize > MAXWIDTH )
  {
      fprintf(stderr,"width > maximum, recompile program\n");
      exit(0);
  }

  /* process the entire image */
  for(i=0;i<ysize;i++)
  {
      for(j=0;j<xsize;j++)
      {
            one_line[j]=getchar();
            putchar( one_line[j] );
            putchar( one_line[j] );
      }
      for(j=0;j<xsize;j++)
      {
            putchar( one_line[j] );
            putchar( one_line[j] );
      }
  }
}
```

```
/*
 *   rotate90.c
 *        - rotate a small image file clockwise 90 degrees.
 *
 *   by A. Kinsman
 *
 */
#include <stdio.h>

#define MAXSIZE 1024

char image [MAXSIZE][MAXSIZE];

main( argc, argv)
int    argc;
char   *argv[];
{
  int i,j,size;

  if (argc != 2)
  {
          printf("usage is    rotate90 x&y-dimension <inputfile >outputfile\n");
          exit(1);
  }

  size = (int) strtol(*++argv, (char **) NULL, 0);

      /* read in image */
  for(i=0;i<size;i++)
     for(j=0;j<size;j++)
          image[i][j]=getchar();

      /* write out image */
  for(j=0;j<size;j++)
     for(i=size-1;i>=0;i--)
          putchar( image[i][j]);
}
```

```
/*
 * smoothit.c
 *      - correct each pixel as an average of it's neighbors.
 *
 *   It would be nice to perform this function in the east-west
 *   and north-south direction.  To obtain smoothing in both
 *   directions use rotate90.c and repeat this filter on the
 *   elevation map.  Then use rotate90.c 3 more times to restore
 *   the original orientation.
 *
 *   by A. Kinsman
 */
#include <stdio.h>

#define MAXWIDTH  4096

static unsigned char one_line[MAXWIDTH];

main(argc, argv)
int   argc;
char *argv[];
{
  int val,i,j,xsize,ysize;

  if (argc != 3)
  {
        printf("usage is   smoothit x-size y-size <inputfile >outputfile\n");
        exit(1);
  }

  xsize= (int) strtol(*++argv, (char **) NULL, 0);
  ysize= (int) strtol(*++argv, (char **) NULL, 0);

  if( xsize > MAXWIDTH )
  {
      fprintf(stderr,"width > maximum, recompile program\n");
      exit(0);
  }

  /* work on all lines */
  for(i=0;i<ysize;i++)
  {
      /* get an entire line */
      for(j=0;j<xsize;j++)
            one_line[j]=getchar();

      /* adjust an entire line, except first and last pixel */
      for(j=1;j<xsize-1;j++)
            one_line[j]= (one_line[j-1] + one_line[j+1])/2;

      /* output and entire line */
      for(j=0;j<xsize;j++)
            putchar( one_line[j] );

  }
}
```

Appendix D   PostScript Stereogram Programs

```
%!
%   IMAGETOP.PS
%      - preamble to render a PostScript image
%
%  by A. Kinsman
%
/width  640 def        % adjust these to the size of
/height 200 def        % your image.

/inch { 72 mul } def

/picstr width string def

8.0 inch 8.0 inch scale    % x and y scaling size

% UNCOMMENT to make it whiter.
% { 1.2 mul } settransfer

% UNCOMMENT to threshold it.
% { .45 lt { 0 } { 1 } ifelse } settransfer

% UNCOMMENT to reverse black&white.
% { 1.0 exch sub } settransfer

% UNCOMMENT to reverse white and black and color logrithmically.
% { 1.0 add log 3.321 mul 1.0 exch sub } settransfer

width height 8
[width 0 0 height neg 0 height]
{currentfile
 picstr readhexstring pop}
image

( INSERT ASCII HEX VALUES HERE, one per pixel.          )
( DO NOT PLACE COMMENTS BETWEEN THE "image" operator )
(   and the image data.                               )
(    Eample-          00 01 02 03 AA BB CC DD EE FF   )
(        represents only 10 pixels worth of data      )

showpage
```

```
%!
%   DOTOP.PS
%       - generate a random dot stereogram from a
%         collection of elevation data points.
%   By  A. Kinsman
%   first some definitions
/inch { 72 mul } def        % points per inch in printing world
/width   240 def            % number of elevation points per line
/height  240 def            % number of lines of elevation points
                        % maximum layers in resulting stereogram
/max-elev width 20 div cvi def
                            % offset between alignment sights
                        %    must always be greater than max-elev
/offset width 4 div cvi dup max-elev lt { pop max-elev } if def

/xscale 8.30    def                % x direction scale size (inches)
/yscale { xscale 5 div 4 mul } def % y direction scale size (inches)

%
%   FOR DEBUG flip this boolean
/random-dot   true  def      % set to false for a gray-scale image
/laser-print true  def      % set to false for on-glass PC display

%
%   each line is composed in this array as it is displayed
/elev-vals width string def             % elevation values
/pict-line width offset add string def  % random dot values

%
%   assume a printer with 8.5x11 inch paper in portrait mode
%   leave a 1/2 inch at top of picture.
laser-print true eq
{ 8.5 xscale sub 2 div inch
  11.0 yscale sub .5 sub inch translate      }  % for laser printer
{ 2.5  inch  2.0    inch translate           }  % for PC display
ifelse

%
%   apply sighting marks and copyright notice
/site-mark { 6 14 rlineto -12 0 rlineto closepath fill } def
random-dot true eq
{
  xscale dup 2 div exch offset mul width offset add div 2 div sub inch
  yscale inch moveto site-mark        % left one
  xscale dup 2 div exch offset mul width offset add div 2 div add inch
  yscale inch moveto site-mark        % right one

  % copyright notice - adjust to your name
  /Symbol findfont 12 scalefont setfont
  .2 inch -.2 inch moveto (\343) show
  /Helvetica findfont 10 scalefont setfont
  .4 inch -.2 inch moveto (A. Kinsman 1992) show
}
if

%
%   scale up for image command
xscale inch yscale inch scale     % use most of whole page

%
%   convert a line of elevation values ( 0->x'FF' ) to the desired
%   dot value, white or black.  Scaled between 0 and max-elev.
/randot-one-line {
        currentfile width string readhexstring pop  % get the data in elev-vals
```

```
        /elev-vals exch def

    0 1 width offset add 1 sub {                % process entire line

            % beginning of new line?  produce a random pixel.
        dup offset lt                           % i - i bool
          { pict-line exch
              rand srand rand          % increase randomness
            1 and 0 eq {    0 put }
                  { 255 put } ifelse
          }

          % not beginning of line,  convert an elevation point.
          { dup offset sub             % - i (i-offset)

              % if off right edge elevation is zero
              % otherwise scale it
            dup width lt               % - i (i-offset) bool
            { dup elev-vals exch get   % - i (i-offset) elev
              % 255 exch sub           % UNCOMMENT to invert altitude
              max-elev mul 255 div cvi % - i (i-offset) elev'
            }
            { 0 }                      % - i (i-offset) 0
            ifelse
                                       % get color for this point
            add                        % - i (i-offset+elev')
            pict-line exch get         % - i pixel-color[elev']
            pict-line 3 1 roll         % - string i pixel-color[elev']
              put                      % -
          }
          ifelse
      } for
      pict-line                % must return this one
} def

%
% produce a gray-scale plot of data or random dot stereogram
random-dot true eq
{                            % for random dot stereogram
    width offset add height 8
    [width offset add 0 0 height neg 0 height]
    { randot-one-line }
}
{                                % for gray-scale elevation map
    % { 1.0 exch sub } settransfer   % UNCOMMENT to invert video
    width height 8
    [width 0 0 height neg 0 height]
    { currentfile elev-vals readhexstring pop }
}
ifelse
%
%  Elevation data points, there are width*height of them in
%  ascii-hex format.  Immediately following the "image" operator.
%  They are expected to be from 0->x'FF' as they are scaled by
%  the code above to a value set in preamble to "MAX-ELEV"
image
( INSERT ASCII HEX VALUES HERE, one per elevation        )
( DO NOT PLACE COMMENTS BETWEEN THE "image" operator )
(   and the image data.                              )
(   Eample-            00 01 02 03 AA BB CC DD EE FF  )
(       represents only 10 points worth of data         )
showpage
```

```
%!
%    BITOP.PS
%      -render a random dot bitmap image in PostScript
%
%    By A. Kinsman
%
/inch { 72 mul } def

/width  1000 def      % width  of image
/height 800 def       % height of image
/offset width 4 div def % offset between sight angles, most often

/xscale { 8.3                    inch } def   % x dir scaling
/yscale { xscale 5 div 4 mul        } def   % y dir scaling

% allocate space for scanline of input
/picstr width 7 add 8 idiv string def

%
%   assume a printer with 8.5x11 inch paper in portrait mode
%   leave a 1/2 inch border at top of picture.
8.5 inch xscale sub 2 div
11.0 inch yscale sub .5 inch sub translate

%
%   apply sighting marks
/site-mark { 6 14 rlineto -12 0 rlineto closepath fill } def
xscale dup 2 div exch offset mul width offset add div 2 div sub
yscale moveto site-mark          % left one
xscale dup 2 div exch offset mul width offset add div 2 div add
yscale moveto site-mark          % right one

% copyright notice - adjust to your name
/Symbol findfont 12 scalefont setfont
 .2 inch -.2 inch moveto (\343) show
/Helvetica findfont 10 scalefont setfont
 .4 inch -.2 inch moveto (A. Kinsman 1992) show

%
%   scale up for image command
xscale yscale scale    % use most of whole page

width height 1
[width 0 0 height neg 0 height]
{currentfile
 picstr readhexstring pop}
image

( INSERT ASCII HEX VALUES HERE,                          )
( DO NOT PLACE COMMENTS BETWEEN THE "image" operator )
(    and the image data.                              )
(    Eample-            00 01 02 03 AA BB CC DD EE FF  )
(       represents only 80 pixels worth of data       )


showpage
```

```
%!
%  MANDELBROT.PS
%
%  A. kinsman
%  6/11/92 computing a PS image with an equation.
%  In this example...the Mandelbrot set.
%
%  some formalities
/inch { 72 mul } def


%
%  get the following values from pressing the tab
%  key on your fractint package on your PC.
%/xmin  -2.0 def       % most of the set
%/xmax   1.0 def
%/ymin  -1.2 def
%/ymax   1.2 def

/xmin  -.4269 def     % interesting place 1
/xmax   .1662 def
/ymin -1.0553 def
/ymax  -.6106 def

%/xmin  -.8490 def     % interesting place 2
%/xmax  -.8270 def
%/ymin  -.2330 def
%/ymax  -.2180 def

%/xmin  -.1121 def     % interesting place 3
%/xmax  -.0834 def
%/ymin  -.8529 def
%/ymax  -.8318 def


%
%  Number of points plotted in the x and y dimensions
%
%  Make these large and it could take forever.  Your
%  fellow workers will get tired of you monopolizing the printer!
%  Niter is effectively the number of gray levels printed.  Again
%  the device loops this many times for each pixel, so keep it
%  down or run the job overnight.
/xpoints  500 def
/ypoints  500 def
/niter     20 def      % number of iterations.


%
%  routine that produces points for the image command
%  The return value must be a single character in a string
%  on the stack to satisfy the 8 bit image command that calls
%  this routine.
/xsize xmin xmax sub abs def
/ysize ymin ymax sub abs def

/x      0 def
/y      0 def
/str  ( ) def

/getpoint {

        x xsize mul xpoints div xmin add /cx exch def
        y ysize mul ypoints div ymin add /cy exch def

        /rx      0.0 def
        /ry      0.0 def
```

```
      0      % iteration counter on stack
      % typical Mandelbrot loop
      {
            1 add          % increment counter on stack
            rx dup mul ry dup mul sub /xtemp exch def
            2.0 rx mul ry mul cy add  /ry exch def
            xtemp cx add /rx exch def
            dup niter eq { exit } if
            rx dup mul ry dup mul add 4.0 gt { exit } if
      } loop

      % make return string with first and only char
      % = stack value of 0->255
      255 niter div mul cvi
      str 0 3 -1 roll put str

      % increment x coord
      x 1 add /x exch def

      % new line test? increment Y coord, and reset x
      x xpoints ge { /x 0 def y 1 add /y exch def } if
} def

%
% print it landscape on a 300 DPI printer, use whole page, 1/2" border
8.0 inch .5 inch translate
90 rotate
10.0 inch 7.5 inch scale

%
%  use the image command for nice half toning, default to
%  standard screen angle and frequency.
xpoints ypoints 8
[xpoints 0 0 ypoints neg 0 ypoints]
{ getpoint }
image

%
%  print the page
showpage
```

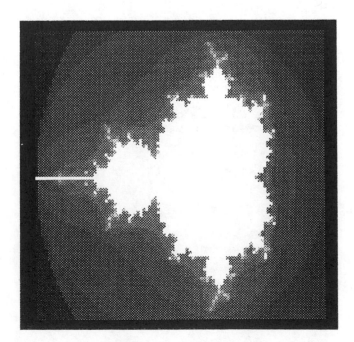

Appendix E   Gray-Scales of the Stereograms

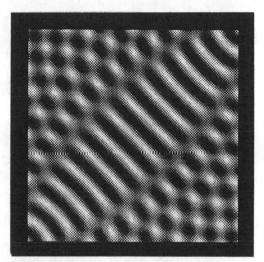

Pond - page 42.

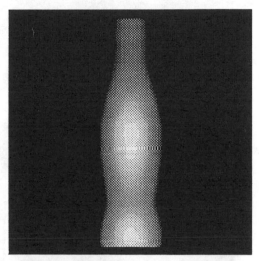

Coke bottle - page 43.

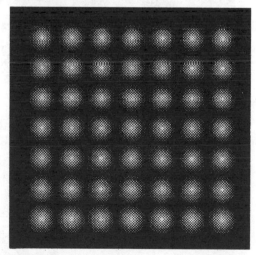

Cones - page 44.

Vase - page 45.

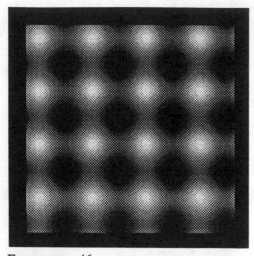

Foam - page 46.

Lamp - page 47.

Sine wave - page 48.

Tree - page 49.

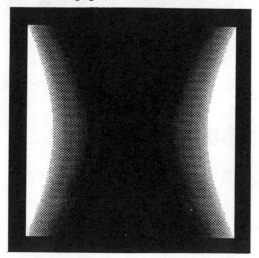

Hyperbola - page 50.

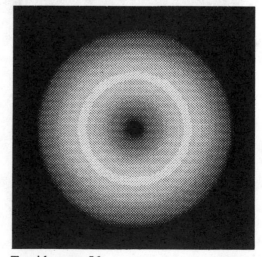

Toroid - page 56.

Beam Splitter - page 58.

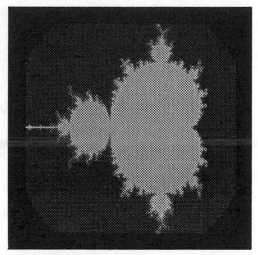

Mandelbrot1 - page 59.

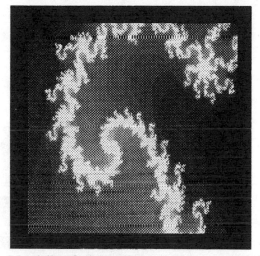

Mandelbrot2 - page 60.

Mandelbrot3 - page 61.

Truchet curve - page 65.

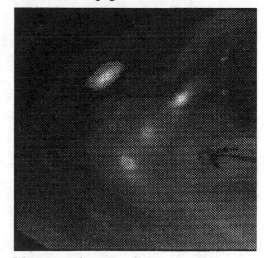

Mars zoom-in - page 67.

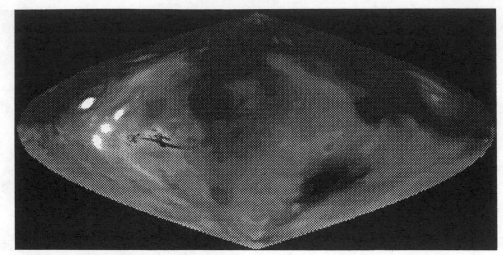

Mars - page 66.

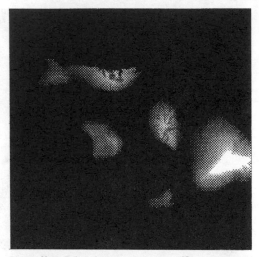

Hawaiian Islands group - page 68.

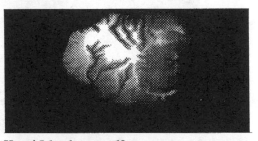

Kauai Island - page 69.

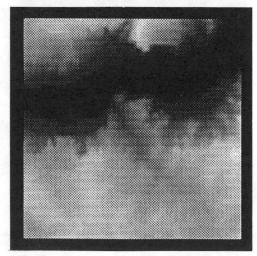

Yosemite Valley - page 70.

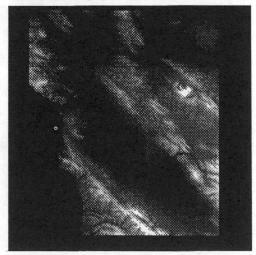

San Francisco Bay - page 71.

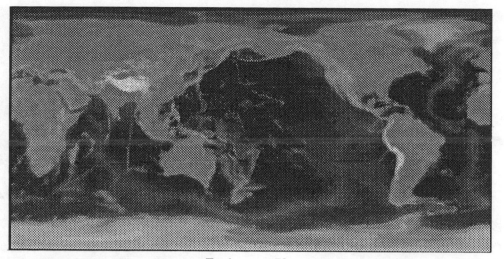

Earth - page 72.

WOW - page 73.

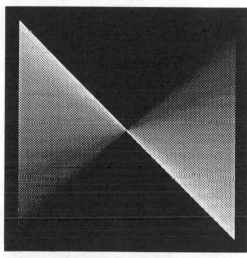

Strings1 - page 74.

Happy Birthday - page 75.

Star - page 76.

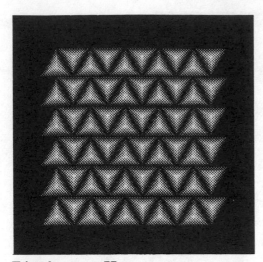

Triangles - page 77.

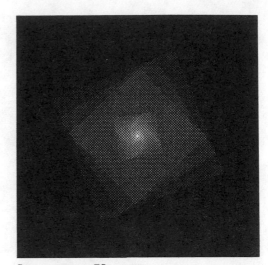

Square - page 78.

Strings2 - page 79.

Strings3 - page 80.

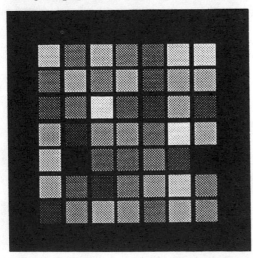

City - page 81.

Propeller7 - page 82.

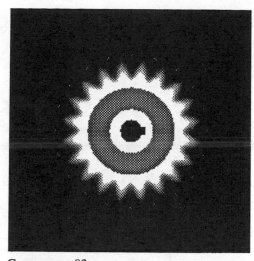

Gear - page 83.

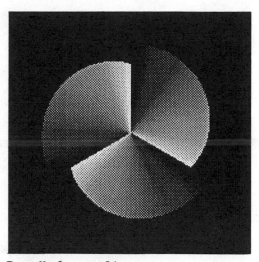

Propeller3 - page 84.

Archimedes2 - page 85.

Chair weave - page 86.

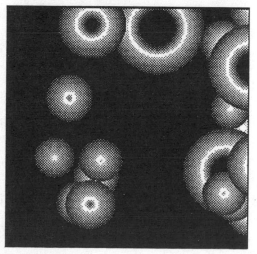

Craters - page 87.

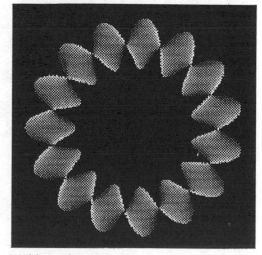

Mobius strip - page 88.

Snake skin - page 89.

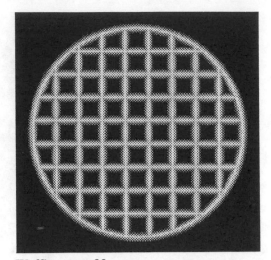

Waffle - page 90.

Clouds - page 91.

Hubcap - page 92.

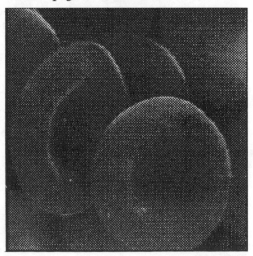

Blood cell - page 94.

Me - page 95.

Glossary and Acronyms

anaglyph   A stereo image where each eye views the page through a filter, usually green glass over the left eye, and red over the right. The two stereo images are printed on top of each other. The image for the left eye is in red ink, the image for the right eye in green ink.

ASCII   American Standard Code for Information Interchange. A 7-bit code commonly used by computers for storing and transferring text.

autostereogram   Refers to viewing of stereoscopic images without the aid of special head gear. For this reason random dot stereograms are often termed random dot autostereograms.

CAD   Computer Aided Design.

DIRDS   Double Image Random Dot Stereogram. Two images side by side are fused to form the stereoscopic view.

DPI   Dots Per Inch.

FAX   Facsimile transmission.

lenticular   Long vertical cylindrical lenses placed over a photograph that restrict the view of each eye to a different portion of the image. A common way of producing thin, portable, and full color stereoscopic images.

pixel   Picture element. Sometimes used synonymously with an elevation point in the elevation grids, primarily because DOTOP.PS permits viewing of the elevation maps both as images and stereograms.

PPI   Pixels Per Inch.

RDA   Random Dot Autostereogram.

RDS   Random Dot Stereogram.

RIP   Raster Image Processor.

RLA   Random Line Autostereogram.

RLS   Random Line Stereogram.

SIRDS   Single Image Random Dot Stereogram.

Acknowledgments

Special thanks go to Edward M. Kinsman for convincing the author that this book needed to be written, for prodding me into completing it in a timely fashion, and for publishing it.

Special thanks go to Susan Pohl of Ithaca, New York, for editing this book.

My admiration and thanks go to the following contributors of illustrations:

Greg Alt
  email:"galt%peruvian@cs.utah.edu"
  for permission to include the STEREOGRAM.C program coded as a stereogram, with its human readable equivalent.

Brian T. Carcich
  email: carcich@moc.tn.cornell.edu
  for assisting me in locating the 1440x720 Mars elevation map.

Ian Collier
  email: Ian.Collier@prg.ox.ac.uk
  for permission to include the "CUBE and SPHERE" stereogram.

Mike Hallesy
  email: hal@kpc.com
  for assisting me in locating the 205x205 Yosemite Valley elevation map.

Robert Michaels
  of Rochester Photonics Corporation
  for permission to include the 1-to-9 laser beam splitter as both a stereogram and a gray-scale image.

Robert Scott &
Gordon Flanagan
  email: Robert.Scott@newcastle.ac.uk
  email: G.J.Flanagan@newcastle.ac.uk
  for permission to include many of their stereograms and for inspiring the two directional RDS on page 40.

Martin D. Simon
  email: msimon@physics.ucla.edu
  for permission to include several of his stereograms.

USGS
  United States Geological Survey
  EROS Data Center
  Sioux Falls, SD 57198
  for compiling the data sets that enabled generation of the stereograms of:
  Hawaii, San Francisco, Mars, Earth, and Yosemite Valley.

Perceptron, Inc.
  Perceptron, Inc.
  23855 Research Drive
  Farmington Hills, MI 48335
  for providing the database that R. Scott & G.J. Flanagan used to produce "Superhuman".

References

Julesz, B. 1971.

Foundations of Cyclopean Perception.
Chicago: Univ. of Chicago Press.

Kahn, D. 1967.

The Codebreakers.
New York: Macmillan Co.

Kontsevich, L. L. 1986.

An Ambiguous Random-Dot Stereogram Which Permits Continuous
Changing of Interpretation, Vision Research, Vol. 26, No. 3,
pp. 517-519.

Ninio, J. and Herlin, I. 1988.

Speed and Accuracy of 3D Interpretation of Linear Stereograms,
Vision Research, Vol. 28, No. 11, pp. 1223-1233.

Slinker, G. S. and Burton, R. P. 1992.

The Generation and Animation of Random Dot and Random Line
Autostereograms, Journal of Imaging Science and Technology,
Vol 36, No. 3, May/June 1992, pp. 260-267.